Economy in Romania and the Need for Optimization of Agricultural Production Structures

Jean Vasile Andrei / Gheorghe H. Popescu

Economy in Romania and the Need for Optimization of Agricultural Production Structures

PL ACADEMIC RESEARCH

Bibliographic Information published by the Deutsche Nationalbibliothek
The Deutsche Nationalbibliothek lists this publication in the Deutsche Nationalbibliografie; detailed bibliographic data is available in the internet at http://dnb.d-nb.de.

Library of Congress Cataloging-in-Publication Data
Jean-Vasile, Andrei, 1982- author.
Economy in Romania and the need for optimization of agricultural production structures / Jean Vasile Andrei, Gheorghe H. Popescu. —First edition.
pages cm
Includes bibliographical references.
ISBN 978-3-631-65826-0
1. Agriculture—Economic aspects—Romania. 2. Agricultural productivity—Romania.
I. Popescu, Gheorghe H., author. II. Title.
HD2052.J43 2014
338.109498—dc23
2014039080

Cover Image:
Champs de blé à l'orée de la forêt (quartier de la Robertsau, Strasbourg)
http://commons.wikimedia.org/wiki/File:Wheat_field_to_forest_%28Roberstsau,_Strasbourg%29.jpg

ISBN 978-3-631-65826-0 (Print)
E-ISBN 978-3-653-05012-7 (E-Book)
DOI 10.3726/978-3-653-05012-7

PL Academic Research is an Imprint of Peter Lang GmbH.

Peter Lang – Frankfurt am Main · Bern · Bruxelles · New York · Oxford · Warszawa · Wien

This publication has been peer reviewed.

www.peterlang.com

Table of Content

Introduction 7

Chapter I
Importance of Agriculture in Romanian Economy....................11

1.1 General agricultural context of the Romanian economy................11
1.2 Comparative developments of Romanian
agriculture vs. European agricultural model 23

Chapter II
Literature Review and Model Settings Research........................29

2.1 General remarks..29
2.2 The Research objectives 32
2.3 Theoretical Framework 33
2.3.1 The Research characteristics 36
2.4 Forecasting economic results for simulated
agricultural modules (total profit and profit
per hectare), 2013–2017...37
2.5 Determination of agricultural modules' budgets..........................39
2.5.1 Underpinning crops' rotation of the
simulated modules 39

Chapter III
Alternatives in Valuing The Agricultural Potential....................41

3.1 Designing a farming module with 60 ha surface,
in lowland agricultural area 41
3.2 Designing a farming module with 100 ha surface,
in lowland agricultural area 57
3.3 Designing a 250 hectare farming module,
in lowland agricultural area 71
3.4 Forecasting a module of farming to an area
of 500 ha, in lowland agricultural area 85
3.5 Forecasting a module of farming to an area
of 800 ha, in lowland agricultural area 98

Conclusions 113

Bibliography 121

Appendixes 129

Introduction

To identify solutions meant to value the agricultural potential under optimal conditions as one of the fundamental issues in emphasizing the role of agriculture in the context of national economy by means of improving the results in this field and enhancing stability, predictability and sustainability of agricultural production.

In order to set and carry out a plan for vegetal agricultural production it is necessary to take into account a series of various endogenous and exogenous factors that production depends on, such as the availability and structure of land, crop plans and rotation, the capital and willingness to invest money, the labor force, the demand and supply for agricultural products, the environmental conditions, and not at least the production costs.

Valuing the potential of Romanian agriculture involves a comprehensive effort to develop coherent and perspective strategies by which to identify factors that may contribute to the sustainable development of this sector. Romania's integration in the European Economic Area and its adaptation to the subsequent changing demands imply a redefinition of the entire national agricultural sector, but also the ways and means to exploit the available agricultural potential by referencing the appropriate national food policies to Common Agricultural Policy (CAP).

It is necessary to understand the current realities through which the CAP transforms itself, in the last time, in the object of political negotiations between the old and new members, assisting in a reorientation of its agricultural production to rural development, mutation that obviously not all members benefit especially the new states that have a significant, but weak and not adequately capitalized agricultural sector.

Agriculture is a vital sector of the national economy with wide influences on macroeconomic outcomes, contributing significantly to Gross Domestic Product (GDP) and Gross Value Added (GVA), while being a potential source of improbability value of these macroeconomic indicators.

Development of agriculture, over the 23 years of transition has highlighted the major role it has in the national economy, both in terms of contribution to the GDP, but by high levels of self-consumption. Evolution of agricultural GDP reflects statistical reality specific to transition process, the high proportion of this is not based on sustainable growth in agricultural production, but rather sharp decline and collapse of other sectors. Sizing artificial agricultural GDP share defines a specific reality of the Romanian economy, a disarticulation of economic sectors, and agriculture is a strong resonance box of instabilities generated by the mechanisms of formation the GVA in the economy.

Significant reduction in the contribution of agriculture to GDP over the analyzed period is the result of a lack of agricultural policies articulated, characterized largely by the dissolution of agricultural production structure and reduced accumulations in agriculture based on the lack of investment and inadequate use of available capital in this sector. In this respect, the share of agriculture in capital investment in fixed assets is significantly below the level of its contribution to GDP. Assessment of the contribution of agriculture to GDP in the Romanian economy is the result of significant failures of macroeconomic statistics. In fact, we can say that the role of agriculture in achieving GDP is somewhat flawed, and its contribution is far superior level highlighted, even if we take into account the evolution of the share of consumption of agricultural products from own resources, the monthly expenditure of households rural households and the farmers in particular.

Agriculture is the main activity in the rural communities of Romania, representing the only source of livelihood for a significant proportion of the rural population.

The transition to a market economy in Romania and implementation of strategies emphasized the sector reform and downsized national agricultural sector, contributing significantly to the creation and accentuation of serious economic and social imbalances that rural areas had either corrected or absorbed them.

Agriculture significantly transformed under the pressure of reform effects from a significant economic sector in a social shock absorber or rather a safety net for people in rural communities has faced with high unemployment and increased migratory flows from de-industrialized urban areas.

The observed effect represents a significant increase in the degree of ruralization trend. Although this level tends to stabilize in the last period, the share of rural population in total population in not the same of 44.9% after 2007, the value confirms not only the ruralism stressed of the Romanian society, but a distancing from the European socio-economic model.

Agriculture is equally an alternative employment of labor available in the economy, if we take into account that we find about 30% of the country's working population in agriculture. Increasing the use of available labor in the agricultural production mechanism can actively contribute not only to improve the living standards of rural communities, but to reduce urban-rural disparities in Romania.

Equally, agriculture is the main element in supporting self-consumption and subsistence population, ensuring a minimum and precarious level of food security. Agricultural production is mostly geared towards meeting their needs for households and not for sharing and recovery in the market, such as is natural for a functioning market economy. Increasing proportions of counter food consumption highlights the precariousness of the stability and sustainability of the Romanian economy as a whole. Share equivalent value of agricultural products from own resources monthly expenditure of households is doubled by the share of agriculture participation to GDP, which is specific to other European economies in the eighteenth and nineteenth centuries.

Romanian agriculture, despite its economic potential above the average community space, continues to be one of subsistence, low-efficiency that cannot meet the guidelines of convergence with the European agricultural model, distinguishing some incompatibility to it.

Agriculture is a fundamental economic sector in the national economy through the high potential of the resources available in terms of agricultural land, population and traditions of rural communities.

Romanian agriculture is facing a strong systemic imbalance and recent developments in the sector, complemented by price instability and identity crisis of production structures, and scare use of resources.

Although, according to European statistics, agriculture has significant potential (on average approx. 0.41 ha/capita of arable land compared to the European average of 0.212 ha/capita, superior to some countries United Kingdom 0.10 ha/capita, France 0.29 ha/capita, Germany 0.14 ha/capita and Poland 0.30 ha/capita), Romanian agriculture results are modest compared to those registered in the Community.

Lack of high performance in Romanian agriculture is due to a combination of factors, among which we can mention:

- High dependence on natural conditions (rainfall) due to destruction of the irrigation system functionality (in 2011 were only 3.2% irrigated surfaces of total area arranged for irrigation); Poor fitting of equipment and the use of outdated production technologies;
- Allocating insufficient, ineffective and discretionary financial resources without selecting priorities, targets etc.

It can be argued that subsistence agriculture in Romania is a modus vivendi not only for farms, but for the economy, which deepened the gap to the European agricultural model, which is open to market.

Chapter I
Importance of Agriculture in Romanian Economy

1.1 General agricultural context of the Romanian economy

Agriculture represents for Romania a large economic sector with implications in ensuring the national potential development, but also in shaping its own evolving structure models of the national production. The integration in the European Economic space equally determines achieving functional transformation of this sector in the conditions of ensuring a necessary convergence to meet the specific requirements of the European agricultural model. To adap the agricultural highly competitive and functional production structures to the market economy involves not only the reform of agricultural production as a whole, but also a shift in the structure of proprietary production resources.

Reconsidering the role that agricultural sector must hold within the national economy should represent not only the result of a scientific step that most often remains only theoretical and declarative, but to be doubled by incentive measures to increase exploitation degree of agricultural potential that Romania has. In this context, agriculture must be analyzed not only from the perspective of achieving GDP, but from a broader one, as a support for developing Romanian rural area and also as a provider of resources and raw materials for downstream economic sectors and beneficiary of imputs of various industries (fertilizer and amendaments, equipment and automation, etc).

Although in the specialized literature there is a variety of opinions on the subject, the approaches are not uniform, but they vary depending on the position an author chooses to take. Starting from the harsh premise according to which we have inherited *an agriculture born from collectivist delusions and megalomaniac hallucinations, but, anyway, an articulated agriculture in its way* (Dropu, 2007)[1], we can estimate the evolution of

1 Constantin Dropu (2007), *Agricultura şi politicianismul. Un secol de politică agrară în România 1907–2007*, Editura Sedcom Libris, Iaşi

Romanian agriculture has been deeply influenced by political decisions adopted in the last two decades of market economy in Romania.

The evolution of Romanian economy and society in the last 24 years has imposed agriculture as one of the most important branches of economy, with broad implications not only for rural communities as it turns into an important factor in causing and promoting economic growth through multiple implications it has in providing the rural income and ensuring consumption, by creating jobs, most often temporary ones, contributing significantly to the capitalization of local resources and their involvement in the creation of value added.

As it is said even in some official documents (Ministry of European Funds, 2013) the Romanian agriculture sector has a significant growth potential but unfortunately that is not operated properly thus the revitalization of the rural economy and the reorganization of the agricultural sector can represent decisive means in the future development of Romania.

The appearence and development of a new concept of multifunctional agriculture, as the specialized studies show, (Renting et al., 2009) *have won a more and more important role in scientific and political debates concerning the future of agriculture and rural development*, imposed a major shift in terms of the role and place of agriculture in rural communities, providing a new starting framework for agricultural potential, especially in rural areas.

In this broader context, beyond the obvious connection between agriculture and rural communities, we can identify an extensive transformation of the European model, which is applied to the rural communities in Romania, too. As Voicu and Voicu (2006) remark *the model of contemporary European village is far from what the traditional countryside represented in old times. Today, in the European countries, the industry and especially the services have replaced the old dominant activities as farming and livestock.*"[2]

From this perspective, the rural communities have been significantly influenced by the transformations and mutations required for their new adaptation to the European agricultural model, where peasants apparently no longer hold the dominant role in the countryside but as the same authors, Voicu and Voicu (2006), remark *farmers are beginning to be*

2 Mălina Voicu, Bogdan Voicu, *Satul Românesc pe drumul către Europa (The Way of Romanian Village to Europe)*, Editura Polirom, Iaşi, 2006, p. 16

rather a rare phenomenon due to the proportion number they have among working population.[3]

Thus, agriculture imposes a trajectory to local communities in ensuring the development of the rural area, as agriculture represents in fact the main occupation of the inhabitants, not only through the development of livelihoods, but especially by allowing non-farm diversification activities, which the Romanian rural communities have begun timidly, but firmly to experiment. As it is argued in some research studies, (Ministry of European Funds, 2013), Romanian agriculture's contribution to GDP has always been high, although it has faced a reduction in the last 15 years while significant endorsement and the oscillation of agricultural production values induces significant variations and instability to GDP.[4]

The nominal values that agriculture contributes for the GDP in Romania are highly relevant in this regard. In Figure 1.1. there is presented the evolution of the value of agricultural GDP vs. total GDP in Romania, in current prices, in the period 2007–2013.

Fig. 1.1 The evolution of agricultural GDP vs. Total GDP, in Romania, 2007–2013

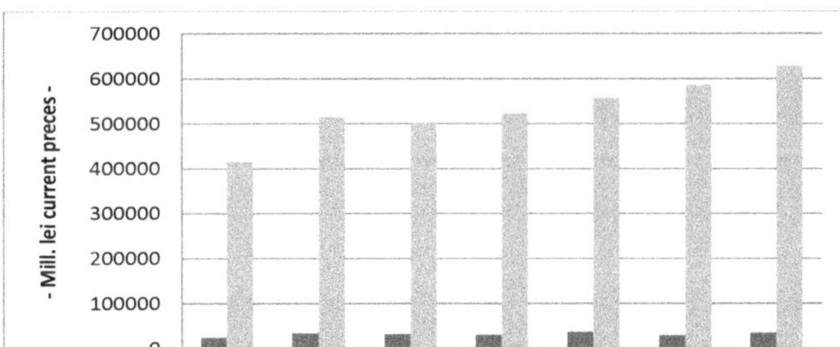

	2007	2008	2009	2010	2011	2012	2013
Agriculture	23992,2	34126,3	32297,8	29874,2	36341,6	28638,1	35192,8
Gross GDP	416007	514700	501139	523693	557348	586750	628581

Source: authors' own computation based on INS-TEMPO online database, 2014

3 *ibidem*, p. 16

4 Ministry of European Funds) (2013), Rezultatele Analizei Documentare. Sectorul Agricultură și Dezvoltare Rurală (The results of the documentary analysis. Agriculture and Rural Development Sector), available at: http://www.fonduri-ue.ro/res/filepicker_users/cd25a597fd-62/2014-2020/Dezbateri%20parteneriale/Rezultatele%20analizei%20documentare/03.06.2013/12.Agriculture%20 and%20rural%20development_22%20mai.pdf, accessed: 21.06.2014

As it can be seen from the data presented in the chart above, both total GDP and agricultural GDP have an upward trend over the entire analyzed period. If in 2007 the agricultural GDP was 23992.2 million lei, representing 5.8% of total GDP, six years later, in 2013, it increased to 11200.6 million lei, respectively +146 68% and amounted to 6.6% of GDP.

As it has been observed from some analyses of the National Bank of Romania[5] in the last decade, the structure of sectors in economy empathized with the Euro Zone (in particular by reducing the share of agriculture), favorable development mitigate possible asymmetric shocks. (NBR 2012).

Evolution of agricultural GDP expressed in nominal values highlights the accentuated dependence of this sector on weather conditions that significantly influence the evolution of agricultural production and a poor equipment necessary to accomplish the capital investment. In this respect it is noteworthy the finding made by the Presidential Commission for Public Policy of Agricultural Development (2013), that argues that in terms of the performance of agricultural and rural development, the real current situation of Romanian agriculture is similar to the one of countries from EU-6 during 1965 to 1970.[6]

In fig 1.2. there is shown the evolution of the agricultural contribution for GDP, in Romania, during 2007–2013.

As it can be seen from the data exposed in the chart above, the contribution of agricultural sector to the GDP in the analyzed period is quite syncope. When Romania joined the EU-27, in 2007, it contributed 5.8% to the GDP, while in 2013 the figure was 6.6%, 0.1% more than in 2011, and in 2012 it recorded the lowest value in terms of this contribution, respectively 4.9%. This trend explains the exposure degree of the agricultural sector to meteorological factors as well as its manual-intensive character.

The relative share the agriculture has had for the GDP is the expression of an agricultural model that is still deep rooted in archaic agricultural practices, where the farmer is still attached to the parcel of land he works it.

5 The National Bank of Romania (2012), *Annual Report 2012, p. 149*

6 Presidential Commission for Public Policy for Agricultural Development (2013), National Strategic Framework for sustainable development of the Romanian agro-food and rural areas during 2014 - 2020 – 2030, available at: http://acad.ro/forumuri/pag_forum_2013_0701Stategie.htm, accessed: 12.03.2014

Fig. 1.2: The evolution of the agricultural contribution for GDP, in Romania, 2007–2013

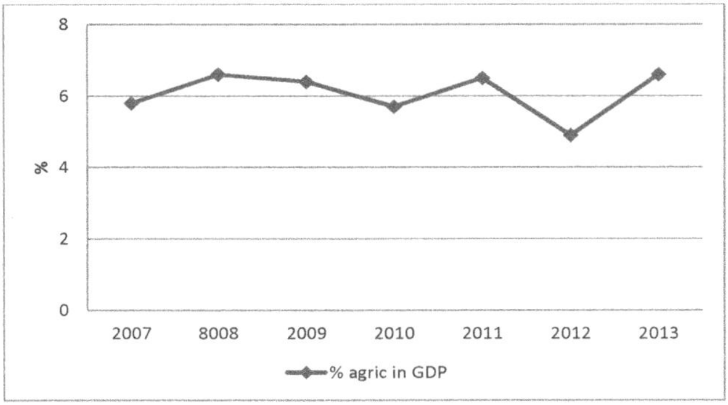

Source: authors' own computation based on INS-TEMPO online database, 2014

To this we add the effects of reforms in the industrial sector where a large part of industrial production capacity was put out of service and, as a result, the contribution of agriculture to added value increased significantly.

If we discuss about the significant influence of the Romanian agriculture in GDP, much higher than in other European economies, the issue is directly connected to the evolution of agriculture productivity in the agriculture sector. For comparison, in fig. 1.3 it is shown the evolution of productivity of the agricultural sector both in the EU-27, and in some Member States, 2009–2011.

Fig. 1.3: The productivity evolution of agriculture sector in some member states EU-27, 2009–2013

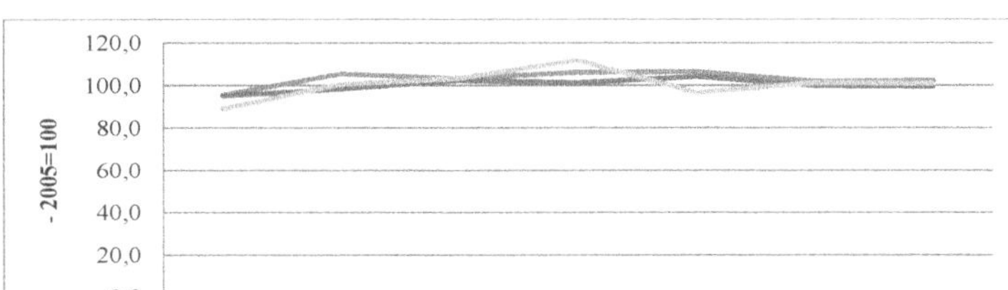

	Bulgaria	Germany	France	Hungary	Poland	România	EU-27
2009	95,5	105,4	103,2	106,1	106,4	101,9	102,2
2010	95,0	98,3	102,3	101,2	104,1	99,7	99,4
2011	89,0	100,4	103,3	111,8	96,6	101,2	100,8

Source: authors' own computation based on xxx

Analyzing the data presented in fig. 1.3. there may be a slight decrease productivity both at EU-27 from 102.2% in 2009 to 100.8% in 2011, and in the countries such as: Bulgaria to 95.5% in 2009 to 89.0% in 2011, Germany from 105.4% in 2009 to 100.4% in 2011, or as applicable Poland, which records the highest reduction in productivity from 106.4% in 2009 to 96.6% in 2011. In Romania, the evolution of this indicator relatively keeps itself around 101%, with a slight decrease to 99.7% in 2010. This consistent trend is a synthetic expression of the fact that, in the case of Romania, agriculture has been and continues to be an economic sector with wide reverberations in the national economy, due to its influences upon a large rural community.

Thus in one of the few studies accomplished by finanaciary institutions in Romania concerning the potential investment in agriculture (BCR, 2011)[7] it is claimed that "the high level of employment and low productivity are the main characteristics of Romanian agriculture; more than 2 million people work in Romanian agriculture, 2.5 times more than in France." (BCR, 2011).

From this perspective, it is interesting to see the evolution of labor used in the Romanian agriculture. In fig 1.4. this development is presented during the period 2000–2013, expressed in annual work units.

Fig. 1.4: The evolution of labor in the Romanian agriculture, 2000–2013

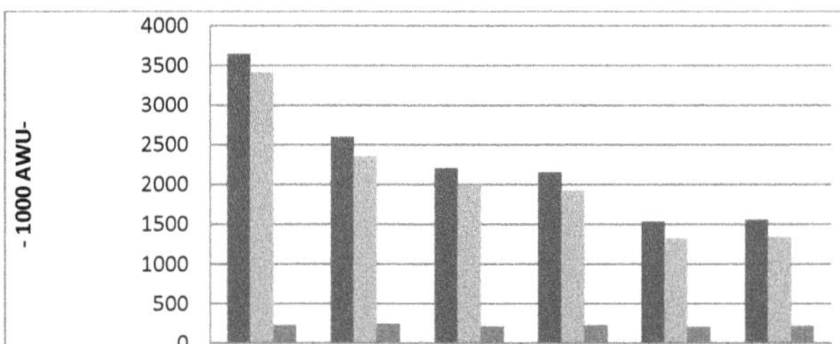

	2000	2005	2007	2009	2011	2013
Total	3645	2596	2205	2152	1532	1560
Unemployed	3415	2353	1994	1925	1326	1340
Employed	230	243	211	227	206	220

Source: authors' own computation based on TEMPO online database, 2014

7 BCR (2011), *Agricultura. Potentialul uriaş al României aşteaptă încă să fie deblocat* (*Agricultura. The huge potential of Romania is still waiting to be deblocked*). Available at: https://www.bcr.ro/ro/.../6a9f7252-8fc7-40f8-9cf8-75274c474e86.pdf, accessed at 05.03.2014

Agriculture involves a significant amount of labor, given that it remains often the only option for a significant share of the Romanian population, where employment opportunities have been already significantly reduced and exacerbated by the economic crisis. In these circumstances, agriculture offers a "safety net" for a numerous part of the population living in rural areas for whom the agriculture means most often the only source of income and provides required livelihood. So if in 2000 there were recorded in agriculture 3.645 million AWU, out of which only 230 000 AWU constitute an employment, adhere only 6.31% of the total workforce in 2013 was only 42.80% of the workforce recorded in 2000.

This drastic reduction in the volume of work was triggered by reducing non-wage category whose values, which decreased from 3.415 million AWU 2000 to 1340000 AWU in 2013, a decrease of 254.85% adhere. Instead employee workforce has remained relatively constant over the entire analyzed period, i.e. from 230 000 AWU in 2000–220 AWU in 2013, which implies a reduction of only 4.55%.

As it is shown in the National Strategy for Employment 2013–2020, in Romania's case, agriculture is oversized as for the employed population in the detriment of other sectors. This happens in the circumstances of a modern economy, while agriculture considerably restricts its employment share in relation to other sectors (Government of Romania. Ministry of Labour, Family and Social Protection for the Elderly, 2014).[8]

Thus, although a significant percentage of the population is employed in agriculture, the evolution of labor productivity shows low values for this indicator of economic efficiency in the agricultural sector. In fig. 1.5 it is presented the evolution of labor productivity in the total economy and in national agriculture during the period 2008–2011.

8 Government of Romania. Ministry of Labor, Family and Social Protection for the Elderly (2014), National Strategy for Employment 2013–2020, available at: http://www.mmuncii.ro/j33/images/Documente/Proiecte_in_dezbatere/2013-08-23/Anexa1_Strategie_Anexe%20I_XII.pdf, accessed at: 12.07.2014

Fig. 1.5: The evolution of labor in agriculture, in Romania, 2008–2011

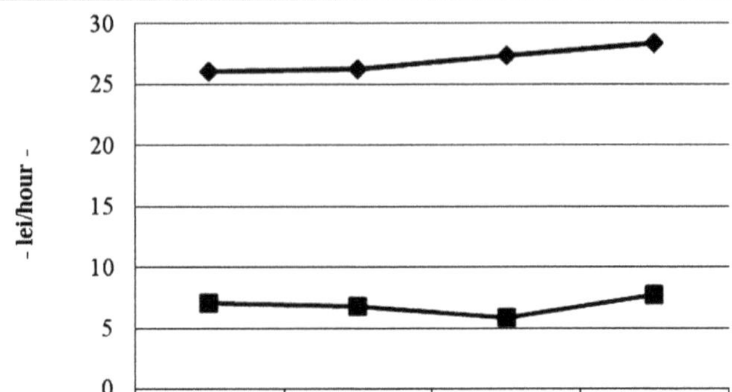

	2008	2009	2010	2011
TOTAL	26.1	26.3	27.4	28.4
Agriculture	7.1	6.8	5.9	7.7

Source: authors's own computation based on TEMPO online database, 2014

In agriculture's case, the labor productivity value expressed in lei/hour has recorded a syncopated evolution over the entire analyzed period. The values of this indicator fluctuates from 26.1 lei/hour in 2008 to 28.4 lei/hour in 2011, thus there is a slight increase of only 8.81% for the entire range without leaps from year to year. In agriculture's case, the situation is quite different with significant variations from a period to another. Only in 2011 this indicator returns to baseline, recorded in 2008, with a slight increase of 0.6 lei/hour repsetiv 7.7 lei/hour after the year 2010 recorded the lowest vaolare the interval, 5.9 lei/time, representing 83.10% of the initial value recorded in 2008.

The evolution of this indicator shows low efficiency of the agricultural sector in Romania recorded by poor equipment and technical equipment of labor. In this context in 2008, the productivity work in agriculture contributes with 27.20% to the productivity recorded in the national agriculture.

The influence of the low level of productivity in agriculture has shown a relative degree of instability to the macroeconomic indicators of the efficiency value recorded in the presented period. Thus in 2009 the productivity of agriculture contributes 25.86% to the national productivity in 2010 with 21.53% and 27.11% in 2011. For comparison in Table 1.1 it is shown the evolution of Regular Farm Labour Force in EU-27, during 2005–2010.

Table 1.1: Evolution of regular farm labour force in some European countries, 2005–2010

Annual working unit

Country	2005	2007	2010
Belgium	66.710	62.630	57.430
Bulgaria	596.620	470.090	389.110
Czech Republic	142.110	133.960	104.600
Denmark	58.320	53.680	50.760
Germany	590.830	555.130	489.660
Estonia	35.850	31.310	24.550
Ireland	148.680	144.330	162.350
Greece	510.520	488.500	372.760
Spain	807.720	790.150	720.860
France	763.240	718.550	692.440
Italy	1.231.890	1.169.430	842.520
Cyprus	25.440	24.280	16.720
Latvia	133.870	103.890	84.640
Lithuania	215.230	176.300	143.820
Hungary	451.990	389.680	402.970
Netherlands	160.570	150.990	141.410
Austria	161.700	159.160	111.180
Poland	2.207.110	2.193.740	1.869.420
Portugal	370.630	315.320	335.780
Romania	2.407.660	2.043.960	1.501.000
Slovenia	90.100	80.020	71.520
Slovakia	95.790	87.480	54.210
Finland	78.660	67.730	55.400
Sweden	67.950	63.370	53.580
United Kingdom	319.030	284.390	246.670

Source: author's based on Eurostat database, 2014

As can be seen from the data presented in Table 1.1. agriculture gathers a major workforce in the European space. It thus appears that in the tradition of European economies, employment in agriculture registers values much lower than in economies recently integrated in the communitary space, such as Romania and Bulgaria.

Although it is remarked a significant reduction of the evolution of regular farm Labour Force, during 2005–2007, the values are relatively significant. If in 2005 regular Labour Farm Force in Romania was 4 times higher than the one recorded in Germany, 3.15 times higher than the one in France and 1.95 times higher than the one in Italy, but comparable to the one in Poland, in 2010, thus even decreasing the values maintained their relative proportion.

Thus, in 2010, regular farm Labour Force in Romania was only 3.06 times higher than the values recorded in Germany, 2.16 times higher than the ones of France. Overall, in the period 2005–2010, regular farm Labor Force in Romania decreased from 2.407660 million AWU to 1.501000 million AWU, 906.660 AWU respectively, which meant that in 2010 regular farm Labor Force in Romania was only 62.34% recorded values in 2005.

An important aspect to bear in mind is the evolution of gross fixed capital, which in the agricultural sector's case, as well in other sectors's case, reflects the interest that investors have for sector development. The labor force, along with the capital form essential elements in increasing exploitation of agricultural potential. In fig. 1.6 it is shown the evolution of gross fixed capital formation in Romania and some European countries in the period 2008–2011.

Fig. 1.6: Evolution of gross fixed capital formation in agriculture in some European countries, 2008–2011

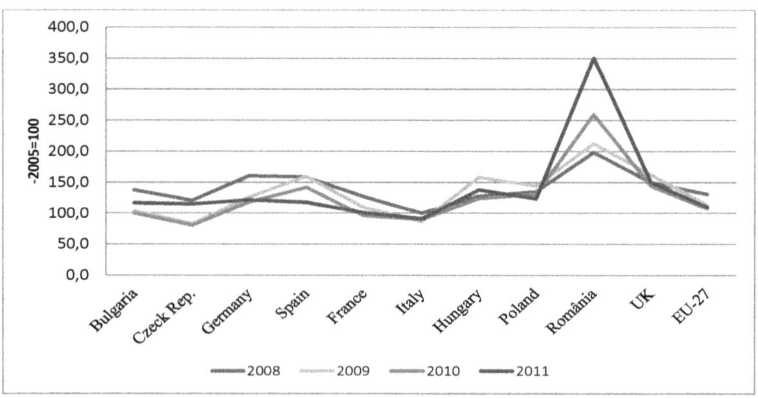

Source: authors' analysis based on Eurostat database, 2014

As it can be observed from the data presented in the chart above, both at the EU-27 level and at most states graphically represented, there is a diminuation in having gross fixed capital formation in agriculture during the analyzed period, with the exception of Romania and Hungary where we notice a consolidation of this indicator.

At the EU-27 level, this indicator decreased from 130.8% in 2008 to 110.1% in 2011, and in Romania's case, there is noticed a significant increase from 198.3% in 2008 to 350.6% in 2011, which means that the national agricultural sector can be seen as an awareness of the role that it has to achieve the national potential economy, but not at its true potential.

Gross fixed capital formation means therefore an increase in the level of equipment with machinery and other farming devices in Romania, but it still remains at a modest level at the values registered in the communitary space. As it is noticed in a study conducted in 2012 (Luke et al., 2012), called "Consolidarea exploataţiilor agricole, performanţele modeste ale agriculturii româneşti" ("Strengthening Farm, Modest Performances of Romanian Agriculture") it can be explained by the scarcity of investment stock and low rates of gross fixed capital formation (showing how much of GDP is invested in the sector and it is likely to contribute significantly to increase competitiveness) (Luca et al., 2012).[9]

In achieving gross fixed capital formation there are some important elements such as: mobilization, development and distribution of investments in economy. The chart below provides the evolution of net investment activities of the national economy during 2008–2012.

As it can be seen from the data presented in Figure 1.7, the net investment in the national economy experienced a significant reduction over the five analyzed years. If in 2008 in Romania there was registered a volume of USD 99525.6 million net investment in 2012 they fell to 9659.1 million lei reaching the value of 89866.5 million lei. In the case of agriculture sector in 2012 there was registered a value approximately equal to that recorded in 2008 and 3362.4 million lei lei compared to 3.3933 billion in 2008. During the period 2009–2010 there was registered the lowest amount of net investment, respectively 2919.5 million lei in 2009 and 2659.8 million lei in 2010.

9 Lucian Luca, Cristina Cionga, Daniela Giurcă (2012), *Consolidarea exploataţiilor agricole (Strengthening farms)*, Bucharest: Ed. Economica, p. 26

Fig. 1.7: Evolution of net investment activities of the national economy in 2008–2012

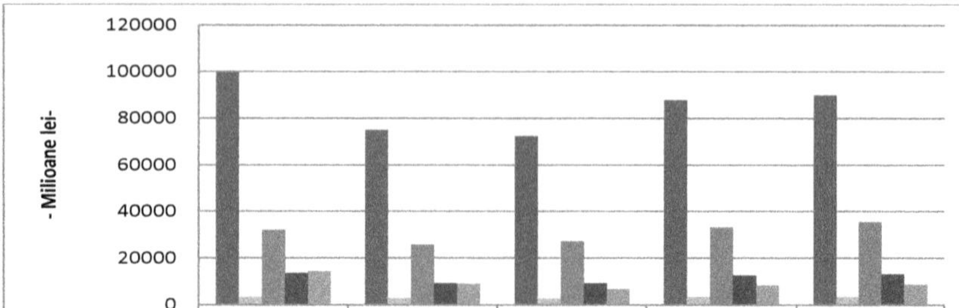

	2008	2009	2010	2011	2012
TOTAL	99525,6	74939,3	72294,7	87815,8	89866,5
Agriculture	3393,3	2919,5	2659,8	3285,1	3362,4
Industry	32071,5	25679,9	27187,7	33108,7	35498,7
Construction	13538	9175,5	9228,3	12649,1	13040,5
Comerce	14403,6	8974,7	6840,9	8389,3	8630,7

Source: authors' own computation based on TEMPO online database, 2014

The evolution of gross value added generated in the agricultural sector is an essential element in demonstrating the weak performances recorded in the agricultural economy of Romania. In figure 1.8 there is shown the evolution of gross value-added at basic prices in some European countries in the period 2008–2011.

Fig. 1.8: Evolution of gross value added in agriculture, in some European countries, 2008–2011

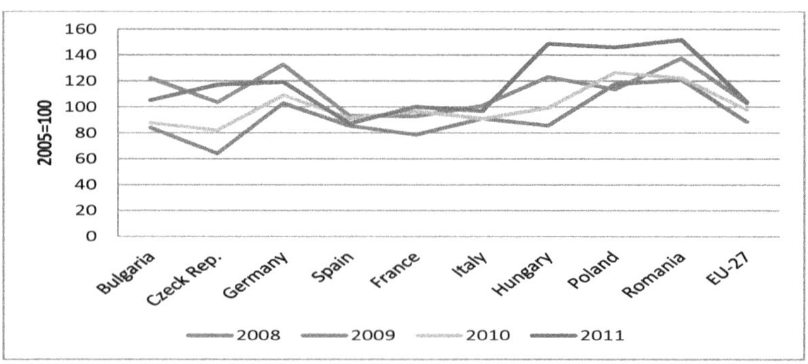

Source: author's based on Eurostat database, 2014

If per whole EU-27 level of this indicator is relatively constant, at around 103% in most European countries analyzed there were significant reductions with the exception of countries like Hungary, from 123% in 2008 to 148.8% in 2011, Romania 137.4% from 151.8% in 2008 to zero in 2008 and Czech car from 103.6% in 2008 to 117% in 2001. Though in the case of Romania the recorded values for this indicator are some significant, they must be analyzed in terms of the role and share of the agricultural sector in the mobilization and use of a significant amount of human and material resources that are not in this context fully rewarded and valued.

Given the findings in the specialized literature[10] that show that precarious investment stock and low rates of gross fixed capital formation (which shows how GPA is invested in the sector) cannot help boost the competitiveness of Romanian agriculture (Luke et al., 2012), it can be said that the Romanian agricultural sector, though in search of performance to come closer to the European agricultural model, it is still far from its values unless somebody understands the importance of this sector in the national economy and Romanian rural communities as a whole.

If in the first part of this chapter we have analyzed the role and place of agriculture in the national economy by comparing it to demands of the European agricultural model this analysis will further consider the evolution of vectors to exploit agricultural potential, namely the evolution of agricultural exploitations in Romania in relation to European space by identifying the main elements of convergence.

1.2 Comparative developments of Romanian agriculture vs. European agricultural model

Understanding the role and place that agriculture plays in realizing the potential that national economy can not be achieved without extensive analysis of the evolution of the utilised agricultural area, given that the earth is one of the fundamental inputs for any of the world's economies. Table 1.2 provides the evolution utilised agricultural area of the EU-27 Member States for the period 2005–2010.

10 Lucian Luca, Cristina Cionga, Daniela Giurcă (2012), *Consolidarea exploatațiilor agricole (Strengthening farms)*, Bucharest: Ed. Economica, p.

Table 1.2: Evolution of the utilised agricultural area, 2005–2010 (ha)

Country	2005	2007	2010
Bulgaria	2.729.390	3.050.740	4.475.530
Czech Rep.	3.557.790	3.518.070	3.483.500
Germany	17.035.220	16.931.900	16.704.040
Spain	24.855.130	24.892.520	23.752.690
France	27.590.940	27.476.930	27.837.290
Italy	12.707.850	12.744.200	12.856.050
Latvia	1.701.680	1.773.840	1.796.290
Lithuania	2.792.040	2.648.950	2.742.560
Hungary	4.266.550	4.228.580	4.686.340
Poland	14.754.880	15.477.190	14.447.290
Portugal	3.679.590	3.472.940	3.668.150
Romania	13.906.700	13.753.050	13.306.130
United Kingdom	15.956.960	16.043.160	16.881.690

Source: authors based on Eurostat datasets 2014

Regarding the evolution of the utilized agricultural area during 2005–2010 there can be noticed a slight reduction in land resources allocated to agricultural operations, excepting Bulgaria that registers the highest increase of utilized agricultural area of 2,729,390 ha year 2005 to 4475530 ha in 2010, which means an absolute increase of utilized agricultural area to 1.74614 million ha (63.98% compared to 2005).

In table 1.3 there is presented the evolution of the changing structure of dairy farms in some EU-27 countries during 2005–2010.

Table 1.3: Evolution of the changing structure of dairy farms in some EU-27 countries, 2005–2010

Countries	Holdings (x 1 000)			Animals (x 1 000)		
	2005	2007	2010	2005	2007	2010
EU-27	2.374	2.119	1.426	18.203	17.680	16.706
Bulgaria	153	121	86	357	350	334
Germany	110	101	90	4.236	4.076	4.165
Spain	42	37	29	1.002	975	910

Countries	Holdings (x 1 000)			Animals (x 1 000)		
	2005	2007	2010	2005	2007	2010
France	104	93	83	3.884	3.815	3.720
Italy	61	63	52	1.860	1.891	1.832
Hungary	16	12	11	287	265	245
Poland	727	651	426	2.854	2.768	2.506
Romania	1.134	1.012	625	1.658	1.587	1.151
United Kingdom	26	28	24	2.065	1.953	1.843

Source: authors' based on European Commission, 2014

As it can be seen from the evolution of the number of farms, presented in the table above, there was a drastic reduction in their number at the EU-27 level, from 2.374 million in 2005 to 1426000 in 2010. Thus, in 2010, at the EU-27 level, there were only 60.6% of all farms registered in 2005. In the case of farms of animals, the reduction is less significant respectively from 18.203 animals in 2005 to 16706 animals in 2010. Moreover the decreasing trend in the number of farms is found at the level of all states analyzed in Table 1.3.

In Romania, there is highlighted a reduction of 509000 farms for the analyzed period. If in 2005 there were 1.134 million farms in Romania, in 2010 they were reduced by 54.90%, registering a total of 625000 farms.

Reducing the number of farms is explained by the generalized tendency to consolidate farms building surfaces across the whole space of EU-27, to which Romania could not make any exceptions. From this perspective the evolution of structure by area farms which they exploit represent a very important perspective. Table 1.4. provides the evolution of agricultural holdings by UAA on class size.

Table 1.4: The evolution of farms in some European countries depending on the used surface, 2003–2010

Country	SAU	2003	2005	2007	2010
Bulgaria	0–5	644,3	511,2	467,8	338,5
	05–10	9,7	10,4	10,1	10,7
	10–20	4,0	4,8	5,5	6,8
	20–50	2,5	2,9	3,6	6,0
	<50	5,1	5,3	6,2	8,4
	Total	**665,6**	**534,6**	**493,2**	**370,5**

Country	SAU	2003	2005	2007	2010
Hungary	0–5 05–10 10–20 20–50 <50	693,3 33,5 21,6 14,2 10,7	641,2 29,0 19,0 14,0 11,7	560,2 24,4 17,2 12,4 12,2	501,6 26,5 19,4 15,4 13,9
	Total	**773,3**	**714,9**	**626,4**	**576,8**
France	0–5 05–10 10–20 20–50 <50	169,5 57,1 63,7 121,3 202,3	147,7 52,2 58,0 109,5 199,8	130,4 48,5 52,2 99,2 197,1	138,8 46,6 50,2 88,5 192,0
	Total	**613,9**	**567,2**	**527,4**	**516,1**
Romania	0–5 05–10 10–20 20–50 0<50	4.205,1 218,9 37,4 9,5 14,1	3.870,7 289,6 65,9 16,1 13,8	3.530,7 300,0 70,1 16,1 14,4	3.593,8 182,4 43,6 17,9 21,2
	Total	**4.485,0**	**4.256,1**	**3.931,3**	**3.859,0**

Source: author's based on European Commission, 2014

The very small farms with 0–5 ha area own the highest percentage in newly integrated economies in the European Economic Area, as is the case of countries such as Romania, Bulgaria and Hungary. In the case of the old Member States, like France, large and very large farms represent a significant share.

In Romania, although it can be noticed a slight tend to reduce the number of small farms with area >5 ha to 4205.1 farms in 2005 to 4205.1 farms in 2010 and an increase of large farms with area <50 14.1 ha of farms in 2005 to 21.2 farms in 2010, the situation is far from increasing leverage of agricultural potential, but rather establish a bipolarity of these farms, where most farms are subsistence with low reliability, and massively exposed to the effects of market volatility.

This situation was proved by the large number of holdings where the holder is manager. The table below provides also the evolution of the number of farms where the owner is the manager at the same time.

Table 1.5: Evolution of the total number of holdings where the holder is manager, 2005–2010

Country	2005	2007	2010
Belgium	47.860	44.120	38.490
Bulgaria	529.050	488.730	364.890
Czech Rep.	39.100	35.850	19.630
Denmark	50.680	43.380	39.910
Germany	360.220	340.830	267.160
Estonia	25.200	20.710	16.950
Ireland	119.080	121.060	135.640
Greece	769.670	796.020	657.860
Spain	874.880	837.140	781.510
France	428.320	383.870	363.810
Italy	1.619.770	1.588.890	1.541.120
Latvia	122.280	104.050	79.470
Lithuania	247.590	223.020	193.590
Hungary	701.060	615.700	564.080
Austria	147.720	134.410	126.310
Poland	2.422.040	2.305.090	1.438.020
Portugal	308.150	258.650	289.110
Romania	4.172.980	3.868.670	3.791.290
Slovenia	77.040	75.210	71.210
Slovakia	65.250	65.840	21.830
Finland	63.800	61.380	56.580
Sweden	66.520	63.470	63.500
United Kingdom	244.470	190.890	162.160

Source: author's based on Eurostat datasets, 2014

Regarding the evolution of the total number of holdings where the holder is manager, out of the data presented in Table 1.5 we can identify a specific situation at the level of the EU-27 Member States. If in the case of countries with an old tradition in the community area we can notice a small number of owners who are the managers of their own farms, the newly integrated states in the EU, excepting Italy, do not follow this rule. In this case, too, Romania stands out with the highest number of holdings where the holder is manager.

In 2005, in Romania, there were registered 4.17298 million of holdings where the holder is manager, in 2010, even if the figure went to 3.79129 million holdings where the holder is manager it continues to be on the first place in the EU-27. Poland and Italy are countries that also recorded a large number of holdings where the holder is manager. In the case of Poland although this number was reduced to 948020 over the period, it continues to be on a position of relative equality with Italy, which in 2010 reached a value of 1.54112 million holdings where the holder is manager to 103,100 in May more than in Poland. Complementary to this development, in the following chart is presented the evolution of national agricultural enterprises according to size classes, 2008–2012.

Fig. 1.9: The evolution of enterprises in national agricultural sector depending on class size, 2008–2012

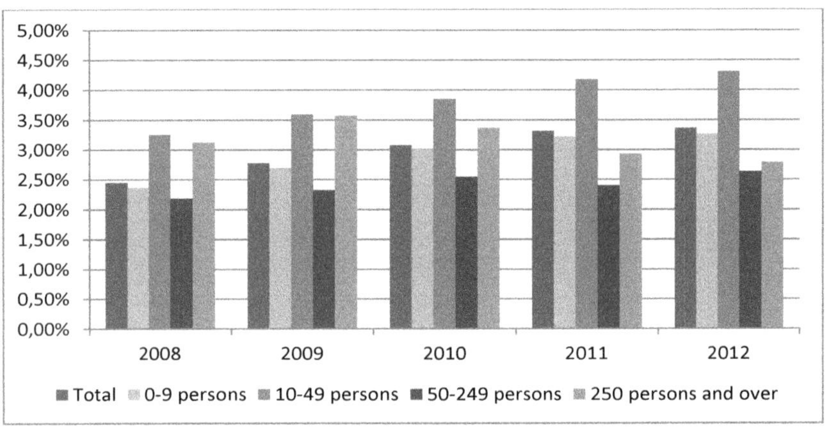

Source: author's own computation based on INS-TEMPO online database, 2014

The evolution of the enterprises from the national agricultural sector in accordance with the class size shows a trend of significant importance in enhancing business argument in the economic field. Thus, if in 2008 the enterprises in the agricultural sector accounted for 2.45% of the total enterprises in Romania, in 2012 they accounted for 3.36% out of the total. In this context, it can be said that the agricultural sector in Romania joins timidly the overall trend imposed by the achieving of the requirements of the European agricultural model.

Chapter II
Literature Review and Model Settings Research

2.1 General remarks

Linear programming is a mathematical method used to achieve production programs, identifying the best result in the context of a system of restrictions identified and undertaken. As a particular expression of mathematical programming through it the objective function is optimized (either maximize or minimize it), under a system of linear equality or inequality restrictions.

Recent studies such as those conducted by Ion and Turek (2012) have used linear programming method to determine the optimal structure of culture by maximizing agricultural profits and declining the expenditure per hectare for farms in the Southeast Region of Development of Romania.

Turek et al. (2011),[11] define a system of production structure optimization considering several endogen and exogenous factors that could influence the achieving of optimal structure of production, such as: farm's environment, market demand, prices of inputs, agricultural technology applied crops rotation and farm's degree of capitalization.

In another study (Subic et al., 2011)[12], linear programming is used for optimal allocation of financial resources if investments in sheep farms.

Drăghici et al. (2010)[13] analyze and present the facilities of using models of linear programming for optimizing production structure of agricultural exploitation by automatic generation of the matrix of economic

---

11 Adrian Turek Rahoveanu, Magdalena Turek Rahovean, Dinu Toma (2011), *Optimization of production structures in order to increase competitiveness of agricultural holdings*, Economics of agriculture, (LVIII): 187–194

12 Jonel Subic, Jean Andrei, Ion Iarca, Dorel Dusmanescu, Ioan Done (2011), *Optimal Allocation of Investment Resources for Agricultural Holding Modernization Using the Linear Programming*, Bulletin of University of Agricultural Sciences and Veterinary Medicine Cluj-Napoca. Horticulture, 68(2): 217–224

13 Manea Drăghici, Livia David, Valentina Tudor (2010), *Facilities of using linear programing for structure optimization of the agricultural holdings activities*, Scientific Papers Management, Economic Engineering in Agriculture and Rural Development, 10(2): 71–75

and technical coefficients, primal-dual solution interpretation and its solving using Simplex algorithm. In another article (Parv and Dzitac 2009),[14] the authors use linear programming to build an optimal multiannual model of agricultural production, considering the objectives of optimal area determination for permanent and annual crops, with the purpose of optimizing farm's revenue.

Using the linear planning meant for the optimization of agricultural production systems has proved to cover a large area also emphasized by the multitude of specialized studies realized during the recent years in this field. Thus have been developed various types of optimization, either for the net results of farms, turnover and profit, or for using the available resources (land, types, fertilizers, mechanical equipment, labor force or irrigation) under the conditions of setting an optimal crop plan. The target functions of these models referred either to the optimal planning of crops rotation under the conditions of maximizing total incomes and aversion to risk (El-Nazer and McCarl, 1986, or Dury et al., 2012)[15], maximizing total incomes earned as a result of planning the whole process of production, cropping, storage and recovering agricultural and animal production (Dobbins et al.,1992)[16], or to planning annual crops under the risk conditions to profitability, ripening, cropping and market demand (Tan şi Comden, 2012),[17] maximizing profit (Nevo et al.,1994,

---

14 Maria Parv, Simona Dzitac (2009), CropPlanWP – *A decision support system for multiannual crop farm planning under probabilistic water availability*, Proceedings of The 44th Croatian & 4th International Symposium on Agriculture, Opatija, Croaţia, p. 237–241

15 Jerome Dury Noémie Schaller, Frédérick Garcia, Arnaud Reynaud, Jacques Eric Bergez (2012), *Models to support cropping plan and crop rotation decisions. A review*, Agronomy for Sustainable Development, 32(2): 567–580 sau Talaat El-Nazer, Bruce McCarl (1986), *The choice of crop rotation: A modeling approach and case study*, American Journal of Agricultural Economics, 127–136.

16 Dobbins, C.L., Preckel, P.V., Han, Y., Doster, D.H. (1992), *An application of linear programming to planning crop systems*, in Proceedings of fourth International Conference on computers in Agricultural Extension Programs, January 28–31, Orlando, FL., p. 376–381

17 Baris Tan, Nihan Comden (2012), *Agricultural planning of annual plants under demand, maturation, harvest, and yield risk*, European Journal of Operational Research, (220): 539–549

or dos Santos et al., 2010),[18] or net profit by means of rigorously planning the operations of repair and crop under the conditions of the availability of production capacity (Higgins et al., 1998),[19] or optimal distribution of field resources (Yagi and Garrod, 2006).[20]

In addition to these researches there are other models of optimization by means of the multiobjective linear planning. Thus, Fritzsch et al. (2011)[21] define a model of four target functions model (maximizing net agricultural production, maximizing net non-agricultural income, maximizing the credit balance of the farm and minimizing the use of labor resources) in order to test the viability of subsistence farms in the South-East and Central Europe and define some public politics meant to improve the condition of those involved in such forms of agricultural organizations.

In Romania the method of linear planning has been used to optimize the agricultural production structures (Ion and Turek, 2012),[22] Dobre et al., 2011),[23] Bran et al., 2011),[24] and recent studies (Ion and Turek, 2012) have

---

18 Nevo A., Oad R., Podmore T. (1994), *An integrated expert system for optimal crop planning*, Agricultural Systems (45): 73–92 sau Lana Mara dos Santos, Alysson Costa, Marcos Arenales, Ricardo Henrique Santos (2010), *Sustainable vegetable crop supply problem*, European Journal of Operational Research, (204) 639–647

19 Higgins A., Antony G., Sandell G., Davies I., Prestwidge D., Andrew B., (2004), *A framework for integrating a complex harvesting and transport system for sugar production*, Agricultural Systems (82): 99–115.

20 Hironori Yagi, Guy Garrod (2006), *An empirical application of the linear programming model to agricultural land-use planning: a valuation of externalities on farmland plots from unused land,* International Journal of Biodiversity Science & Management, (2:3): 227–231

21 Jana Fritzsch, Stefan Wegener, Gertrud Buchenrieder, Jarmila Curtiss, Sergio Gomez y Paloma (2011), *Is there a future for semi-subsistence farm households in Central and southeastern Europe? A multiobjective linear programming approach*, Journal of Policy Modeling, (33): 70–91

22 Ion Raluca Andreea, Turek Rahoveanu Adrian (2012), *Linear Programming in Agriculture: Case Study in Region of Development South-Muntenia*, International Journal of Sustainable Economies Management, 1(1): 51–60

23 Dobre Iuliana, Voicu Radu, Bran Mariana (2011), *The phenomenon of diversifying in production structure and its impact economic, social and ecologic.* Journal Quality – Access to Success, 12(121): 171–175.

24 Bran Mariana, Dobre Iuliana, Stefan Marcela, Voicu Radu (2011), *Opportunities for the vegetal biodiversity - support for the sustainable development in the South Muntenia region, Journal Quality – Access to Success, 12*(121): 32–36.

made use of the linear planning method in order to determine the optimal structure of the crop, by maximizing agricultural profits and reducing the expenditures per hectare for the fields in the South-East area of development in Romania.

Recovering the Romanian agricultural potential under the conditions of market globalization and integration may be possible only by means of optimizing the production structures and harmonizing them with the demands of the national market economy, functional and competitive, which due to improving the level of recovery and distribution of available resources can generate not only high levels of the agricultural production but also the improvement of Romanian agricultural competitiveness on specialized international markets.

In this respect, they proposed the achievement of five distinct modules of vegetal agricultural production, able to supply the recovery of the agricultural potential, under optimal economic and ecological conditions. At the base of simulation for these modules it was taken into account the observance of certain conditions such as:

- the integral cultivation of the agricultural area and lining the established crops to rotation, for each of the five modules of agricultural production;
- carrying out and observing the rotation by imposing some area limits referring to the distribution of each crop in the agricultural production plan;
- lining the expenditures for production per hectare to the limits of the budgets allocated for agricultural production modules;
- the efficient use of the labor resources, capital and technical equipment;
- the achievement of an agricultural production level able to supply a maximum profit for each of the agricultural production module.

They designed five modules of agricultural production, specific to the field area in Romania, having area starting from 60 hectares.

2.2 The Research objectives

The achievement of the research objectives of the present book imposed on one hand the identification of a relevant method to plan the optimal options for the crops within five agricultural areas, proposed as solutions for improving the level of recovering the agricultural conditions and factors and, on the other hand, the base for a theoretical model of research that may be applied under the conditions described and simulated in this research.

The objectives of the research were:

- establishing and defining the optimal production structure for the agricultural
- production model;
- identifying the rotation necessary for the achievement of the optimal production referring to the agricultural production modules and establishing an agricultural crop plan;
- planning the annual and general budget, and estimating them for the year 2017;
- predicting the profit per hectare and module taking into account the production structure determined as optimal, by predicting the sale price and total cost of production between 2013–2017.

2.3 Theoretical Framework

In specialized literature,[25] the general expression of the mathematical model for linear planning use in this research supposes either problems for maximizing or minimizing the target function:

$$f(x) = (x_1, x_2, \dots \dots .., x_n) \qquad (1)$$

where the decisional variables

$$x = [\![(x]\!]_1, x_2, \dots .. x_n) \qquad (1.1)$$

are limited by the restrictions $g_i(x)$

$$g_i(x) = g_i(x_1, x_2, \dots \dots x_n) \leq 0 \qquad \textit{where } i = 1, 2, \dots . m \qquad (1.2)$$

If the target function and the restrictions imposed have a linear character, then the function can be expressed either in the following form:

25 On writing the research model it was taking into account some previous research as: (Cămăşoiu, 1986), (Parv and Dzitac, 2009), (Turek et al., 2011) or (Ion and Turek, 2012)

1) max $f(x)$:

$$\max f(x) = \sum_{k=1}^{n} p_j x_j \tag{1.2}$$

in the following conditions:

$$\sum_{j=1}^{n} a_{ij} x_j \geq b_i \qquad unde\ i = 1, 2, \ldots., m \tag{1.2.1}$$

$$x_j \geq 0 \qquad unde\ i = 1,2, \ldots \ldots, m \tag{1.2.2}$$

2) *or* min $f(x)$:

$$\min f(x) = \sum_{j=1}^{n} c_j x_j \tag{1.3}$$

keeping the correspondent restrictions:

$$\sum_{j=1}^{n} a_{ij}\, x_j \geq b_i \qquad where\ i = 1, 2, \ldots \ldots, m \tag{1.3.1}$$

The restriction in 3.1.1 can be also written as:

$$g_i(x) = \sum_{k=1}^{n} a_{ij}\, x_j - b_i \leq 0 \quad where\ i = 1, 2, \ldots \ldots, m \tag{1.4}$$

The symbols used below have the following meanings:

x_j – the surface held by j crop;
p_j – the profit per hectare of the j crop;
a_{ij} – the consumption from i resource to the area unit of the j crop;
b_i – the volume of the resources of type i;
j – the number of crops;
i – the number of resources.

The general restrictions imposed to the model at 1.3 and 1.4 may be completed with a series of new special restrictions such as:

a) *either superior limitation of the area held by a certain crop (x_l)* in order to observe a series of minimal conditions such as crop rotation, environmental restrictions to a superior recovery of the crop (x_i), imposing the limit:

$$x_i \leq KS \tag{1.5}$$

In which:

K – percent that supplies the minimum demands imposed to the area in order to obtain the crop x_i
S – surface cultivated with xi in the production module.

b) *either inferior limitation of the area held by a certain crop (x_i)* and determining a minimum area of allocated crop in order to improve the chances to obtain a significant profit by imposing the following restriction:

$$x_i \geq K'S \tag{1.6}$$

where:

K' – minimum percent for which the area allocated to crop (x_i) meets the requests imposed to the model

c) *entirely using the agricultural area corresponding to the production model*

$$\sum_{j=1}^{n} x_j = S, \tag{1.7}$$

Where:

S – arable land corresponding to the production model;
x_i – area allocated for crop i;
n – number of crops

In case of this model, there are allocated to S the following values = 60 ha, 100 ha, 250 ha, 500 ha and 800 ha

d) As for microzoning the agricultural production within the agricultural area, restriction (3.7), may be transcript in the form:

$$\sum_{k=1}^{m} x_{ik} = S_i \qquad \text{for } i = 1,2,3 \ldots, \text{n} \tag{1.8}$$

where:
x_{ik} – the area allocated to crop *i*, placed on the plot *k*;
m – number of available plots.

To apply the restriction underlined at point (1.8) can be possible only in case it checks the conditions mentioned at (1.8.1) and (1.8.2), which is:

- the equivalence of the area for the existent plots $[\![(S]\!]_k)$ with the total agricultural area held by the crops $[\![(S]\!]_i)$ of the land

$$\sum_{k=1}^{m} S_k = \sum_{i=1}^{m} S_i \qquad (1.8.1)$$

- the total of the crops planted on the same plot (x_{ik}) must be equal with the area corresponding to that plot (S_k)

$$\sum_{i=1}^{n} x_{ik} = S_k \qquad \text{for } k = 1,2, \ldots, \text{m} \qquad (1.8.2)$$

e) Or, in order to allow for a certain mobility of the solution, in case of applying the restrictions from (1.6) and (1.7), it can be imposed an additional restriction of the type:

$$K_1 S \leq x_z \leq K_2 S \qquad (1.9)$$

Where K_1, K_2 – any percent.

2.3.1 The Research characteristics

The target function of the model takes into account maximizing the profit and identifying the maximum of the gross margin and the level of total expenses in the structure of each simulated production option. In these conditions, the options of the model represent the diversity of agricultural crops possible to be experienced on the existent areas and which meet a complex set of demands imposed for the practical viability of the model.

Certain conditions were imposed to observe the rotation, avoid single-cropping, maximize profit and gross margin as well as minimize the total expenses for each model proposed. After identifying the values for each variable x_i, we present the values of the total profit, total gross margin and total expenses for each option of crop obtained as a result of the optimization plan.

Although the model is based on the demands of a method of planning the agricultural production, it is different from any other models of optimization and planning the agricultural production described in the specialized studies because it focuses on a larger area of application due to the nature of the target functions and their simulation on various production structures corresponding to the proposed modules of crops.

In this respect, starting from the demands imposed from using classical models of optimization, that is linear planning, we designed a multi-objective linear planning model which aims at achieving two maximum functions – in case of profit a total gross margin, and a minimum function – in case of total costs of the farming.

We determined solutions referring to: *the maximum level of the profit corresponding to each production option.*

As a result of running the model we found *target functions of maximizing the profit f(x), minimizing the expenses g(x) and maximizing the gross margin h(x),* which are presented at the end of the five options of the possible production structures, adding the basically option (zero) to each of the five production models. The research can go on based on them by using methods of mathematical modeling that may lead to achieving a future planning of the production structure.

2.4 Forecasting economic results for simulated agricultural modules (total profit and profit per hectare), 2013–2017

Starting with optimal structure of production identified using the method of linear programming, agricultural prices forecasting using method of linear regression[26] and costs' forecasting using the method of average annual growth rate[27]; *the research includes forecasting of profit per*

26 The agricultural products prices are rising in a constant rhythm during 2006–2012 which has imposed choosing a linear function in order to predict the prices during 2013–2017.

27 Predicting the production costs has been made taking into account that the inflation rate has an yarely rhythm equal with 5%

hectare and total for each of the five modules of agricultural production for period 2013–2017.

To forecast economic results in period 2013–2017, prices of agricultural products and costs per hectare have been predicted. In this purpose, an increase of 5% per year has been estimated, as resulted from inflation rate (Appendix 6).

Considering the prices' trend for the products included within the program of production, in the last years 2006–2012, the method of statistical forecast, linear regression, is used. The equation of linear regression for the prices of products from the modules simulated has the following forms:

- Wheat price
  $y = 0.08x + 0.91$ $R^2 = 0.996$;
- Barley price
  $y = 0.068x + 0.77$ $R^2 = 0.996$;
- Corn price
  $y = 0.071x + 0.82$ $R^2 = 0.996$;
- Sunflower price
  $y = 0.12x + 1.363$ $R^2 = 0.996$;
- Sugar beet price
  $y= 0.023x + 0.272$ $R^2 = 0.989$;
- Rape price
  $y = 0.12x + 1.363$ $R^2 = 0.996$;
- Beans price
  $y= 0.28x + 3.186$ $R^2 = 0.997$;
- Soybeans price
  $y = 0.168x + 1.91$ $R^2 = 0.997$.

The values of R^2 closed to 1 demonstrate that the method is valid for estimating the level of price. The prices of agricultural products predicted the estimation of profit per hectare for each crop in the plan of production and for each module of exploitation and the total profit per module are presented in chapter 2.

2.5 Determination of agricultural modules' budgets

The research finalized through the projection of yearly budgets of production for each of the five modules, in the perspective of the period 2013–2017. The process of forecasting the budgets of revenues and expenses for the modules of agricultural exploitations started from the classical underpinning of budgets of each crop from the program of agricultural production, through achieving the estimative calculation regarding the level of costs specific to each input and of net profit, considering the technological spread sheets of each crop. The value of unfinished or in progress production and estimation of inputs' prices for the following year has been added. In this respect, both yearly budgets for each crop from the yearly program of production: wheat, barley, corn, sunflower, rape, sugar beet, soybeans and beans; and multiannual budgets, for the period 2013–2017, for the five simulated modules. The multiannual budgets for modules of 60 ha, 100 ha, 250 ha, 500 ha and 800 ha have been underpinned. The results are presented in Appendixes 18–22.

2.5.1 Underpinning crops' rotation of the simulated modules

Getting the best parameters required to choose a crop rotation system to ensure compliance with environmental requirements, both in the conditions imposed by practicing sustainable agriculture and achieve maximum production attainable under the simulated conditions. In this sense, the choice of design culture was strongly conditioned by the duration, type and applied crop rotation scheme.

Crop rotation is an effective method for optimizing the structure of production in the agricultural production modules. Recently, the specialized literature (Dury et al., 2012) provides a review of patterns for background and design of production plan and crop rotations that can be applied to the farm. To achieve the research objectives, we opted for agricultural production modules proposed for agricultural crop rotation 7 simple sole for modules of 60 ha and 100 ha, respectively with 6 simple sole for holdings of 250 ha, 500 ha and 800 ha, assuming a 5-year crop rotation. Crop rotation scheme includes winter cereals, row and industrial crops and grain legumes; it is presented in Appendixes 1–5

Chapter III
Alternatives in Valuing the Agricultural Potential

3.1 Designing a farming module with 60 ha surface, in lowland agricultural area

Designing an agricultural production structure of 60 ha started from the initial idea according to which, even in the specialized literature were simulated production modules having area of 50 ha (Ion and Turek, 2012)[28] or (Dobre, 2003),[29] the results showed the idea that the area of 60 ha represents the minimum area conditions of the Romanian agricultural farming in order to be economically viable. We have also tried to make sure that this kind of farming may be a relevant solution in the process of aggregation of Romanian land property in order to increase the level of productivity under the conditions of 60 ha projected module that is included in the V class having medium economically potential, reduced viability and which due to simulation in the first year generates a profit rate of 15,5%, thus supplying a profit of 526,3 lei/ha. (Table 2.6)

At the base of this module of agricultural production we took into account the appliance of a five year rotation, with seven simple sole (Appendix 1), specific to the field area, that included both autumn grains (wheat + barley), weeding plants (corn + sunflower), technical plants (rapeseed + sugar beet) but also leguminous vegetable (beans).

To optimize the production structure corresponding to the 60 ha module we started from identifying the initial level of the economic parameters followed in the process of optimization, namely the data referring to the expenses, gross margin a profit per hectare, for each crop, in the first year

---

28 Ion Raluca Andreea, Turek Rahoveanu Adrian (2012), *Linear Programming in Agriculture: Case Study in Region of Development South-Mountenia*, International Journal of Sustainable Economies Management, 1(1): 51–60

29 Iuliana Dobre, *Managementul structurii de producție în exploatațiile agricole*, Editura ASE, Bucureşti, 2003, p. 163

of rotation, according to the budget planned in the initial crop year, 2012, (Appendix 8) and the basic option of the optimization model (Table 3.1).

Table 3.1: Production structure, expenses, gross margin and profit per hectare in initial option, corresponding to 60 ha module

Crop	Area (ha)	Expenses (lei/ha)	Gross Margin (lei/ha)	Profit (lei/ha)
Wheat	13	3073	1000	607
Barley	7	3026	813	374
Corn	17	3161	1150	399
Sunflower	10	2231	805	398
Rapeseed	6	2433	985	647
Sugar Beet	3	6201	1741	479
Beans	4	3041	870	439

Source: based on the rotation in the first year of production

Based on the data presented in the table below, we can see that the reduced levels of expenses per hectare are recorded to sunflower and rapeseed crops: 2231 lei and 2433 lei. Sugar beet, corn and wheat crops have increased gross margins, with a level of over 1000 lei/ha. Increased levels of profit per hectare have been recorded for rapeseed, wheat, sugar beet and beans (between 400 and 600 lei/ha).

The existent production structure leads to the following levels of expenses, gross margin and profit:

$$Ch_0 = \sum_{j=1}^{7} ch_j \cdot x_j = 182543 \text{ lei};$$

$$MB_0 = \sum_{j=1}^{7} mb_j \cdot x_j = 60904 \text{ lei};$$

$$P_0 = \sum_{j=1}^{7} p_j \cdot x_j = 28347 \text{ lei};$$

Where the symbols used have the following significance:
Ch – total expenses;
MB – total gross margin;

P – total profit;
ch_j – expenses on the production unit corresponding to crop j;
mb_j – gross margin on the production unit corresponding to crop j;
p_j – profit on production unit corresponding to crop j;
x_j – area corresponding to crop j.

Variant 1

In the process of optimizing the production structure, in this first option, as a result of applying the linear planning method, we have tried to identify the optimal structure of the crops included in the production module, by means of maximizing the total profit of the farming.

The model corresponding to the no.1 option has the following form:

$$\max(f(x)) = 607\,x_1 + 374\,x_2 + 399\,x_3 + 398\,x_4 + 647\,x_5 + 479\,x_6 + 439\,x_7$$

To this option there are available the following restrictions:

1. $3073\,x_1 + 3026\,x_2 + 3161\,x_3 + 2231\,x_4 + 2433\,x_5 + 6201\,x_6 + 3041\,x_7 \leq 372060$
2. $1000\,x_1 + 813\,x_2 + 1150\,x_3 + 805\,x_4 + 985\,x_5 + 1741\,x_6 + 870\,x_7 \geq 48300$
3. $x_1 + x_2 + x_3 + x_4 + x_5 + x_6 + x_7 + x_8 + x_9 + x_{10} + x_{11} = 60$
4. $x_1 \geq 8{,}5$
5. $x_2 \geq 8{,}5$
6. $x_3 \geq 8{,}5$
7. $x_4 \geq 8{,}5$
8. $x_5 \leq 8{,}5$
9. $x_6 \leq 8{,}5$
10. $x_7 \leq 8{,}5$

The first restriction refers to the level of the expenses for each crop, which summed up must not exceed the 372060 lei maximum allocated budget. This value was obtained by multiplying the maximum expenses per hectare, which according to the data in the table 2.1 correspond to the sugar beet crop, on a total area of 60 ha, starting from the premise that the whole area would be cultivated with the most expensive crop, in the prediction of a pessimistic scenario. The second restriction takes into account the total gross margin that should be greater or at least equal to 48300 lei.

This value was obtained by multiplying the minimum gross margin per hectare which according to the data in the table 4.1 correspond to sunflower crop having the total area of 60 ha, starting from the premise that it would be cultivated the whole area with the crop having the most reduced level of the gross margin per hectare, in the prediction of a pessimistic scenario.

The third restriction refers to the fact that the options, namely the summed up crop areas must not exceed the total area of 60 ha corresponding to the production module. The next restrictions were imposed to the options according to the level of crops insensitivity thus the crops placed on large areas such as wheat, barley, corn and sunflower have inferior limits of 8,5 ha whereas the crops placed on reduced areas such as those of rapeseed, sugar beet and beans have superior limits of 8.5 ha.

The level of 8.5 ha was chosen because it represents the middle area which would be equally allocated to the seven crops. As a result of applying the linear planning method, the following values for the variables x_j and for profit, gross margin and total expenses were obtained:

$$P_1 = \sum_{j=1}^{7} p_j \cdot x_j = 31235 \text{ lei};$$

$$MB_1 = \sum_{j=1}^{7} mj_j \cdot x_j = 57900 \text{ lei};$$

$$Ch_1 = \sum_{j=1}^{11} c_j \cdot x_j = 172131 \text{ lei};$$

Thus, the following production structure for allocating crops is identified:

Wheat: x_1 = 26 ha;
Barley: x_2 = 8,5 ha;
Corn: x_3 = 8,5 ha;
Sunflower: x_4 = 8,5 ha;
Rapeseed: x_5 = 8,5 ha;
Sugar Beet: x_6 = 0 ha;
Beans: x_7 = 0 ha.

In this option of the production structure, the biggest part of the agricultural area is allotted for wheat (43%), and barley, corn, sunflower and rapeseed are cultivated on identical area each of 8,5 ha. Sugar beet and beans crops miss form the production structure.

The condition of maximizing profit is observed, because the profit level is bigger than the basic option, and the area cultivated with rapeseed which is the most profitable crops increases from 6 to 8.5. The area cultivated with wheat which has a big profit per hectare increases as well from 13 ha to 26 ha.

Variant 2

In this option, the target function is unchanged but the restrictions are improved, taking into account the rotation of crops. The model corresponding to this option becomes:

$$\max(f(x)) = 607\, x_1 + 374\, x_2 + 399\, x_3 + 398\, x_4 + 647\, x_5 + 479\, x_6 + 439\, x_7$$

We propose the following restrictions:

1. $3073\, x_1 + 3026\, x_2 + 3161\, x_3 + 2231\, x_4 + 2433\, x_5 + 6201\, x_6 + 3041\, x_7 \leq 372060$
2. $1000\, x_1 + 813\, x_2 + 1150\, x_3 + 805\, x_4 + 985\, x_5 + 1741\, x_6 + 870\, x_7 \geq 48300$
3. $x_1 + x_2 + x_3 + x_4 + x_5 + x_6 + x_7 + x_8 + x_9 + x_{10} + x_{11} = 60$
4. $x_1 \leq 15$
5. $x_2 \geq 8$
6. $x_3 \geq 15$
7. $x_4 \geq 10$
8. $x_5 \leq 8$
9. $x_6 \geq 4$
10. $x_7 \geq 4$

In this option we used restrictions referring to the superior limit of the wheat crop (15 ha) which in the first option was of 26 ha, meaning almost half of the total agricultural area and a limit of the rapeseed crop of 8 ha, which given that it is a plant with big profit per hectare supposed to be allocated the largest area in running the plan.

Restrictions regarding the inferior limit were imposed to barley (8 ha), corn (15 ha) and sunflower (10 ha), as they are crops that even though are placed on large areas in the first option got middle areas of 8.5 ha. Sugar beet and beans crops (that were not present in the first option) were introduced in the production structure due to inferior limitation to the area of at least 4 ha. As a result of applying the method, the following values for the options x_j, profit, gross margin and total expenses are obtained:

$$P_2 = \sum_{j=1}^{7} p_j \cdot x_j = 28482 \text{ lei};$$

$$MB_2 = \sum_{j=1}^{7} mj_j \cdot x_j = 61128 \text{ lei};$$

$$Ch_2 = \sum_{j=1}^{11} c_j \cdot x_j = 184168 \text{ lei}.$$

The following production structure is determined:

Wheat: $x_1 = 11$ ha;
Barley: $x_2 = 8$ ha;
Corn: $x_3 = 15$ ha;
Sunflower: $x_4 = 10$ ha;
Rapeseed: $x_5 = 8$ ha;
Sugar Beet: $x_6 = 4$ ha;
Beans: $x_7 = 4$ ha.

The results obtained in this option show the conditions regarding the rotation of crops, as the production structure includes both grains (wheat, barley, corn) and oleaginous plants (sunflower and rapeseed), technical plants (sugar beet) and leguminous (beans, being considered a good preceding to grains). The area cultivated with wheat is reduced from 26 ha to 11 ha, which means a reduction of the profit compared with option 1.

Restrictions regarding the extension of the area cultivated with wheat but also the recovery of sugar beet and beans crops which are supposed to increase the profit were imposed in the next option.

Variant 3

In the third option, the target function is the same but the restrictions take into account both rotation of crops and their profitability. The model is:

$$\max(f(x)) = 607\, x_1 + 374\, x_2 + 399\, x_3 + 398\, x_4 + 647\, x_5 + 479\, x_6 + 439\, x_7$$

The system of restrictions is:

1. $3073\, x_1 + 3026\, x_2 + 3161\, x_3 + 2231\, x_4 + 2433\, x_5 + 6201\, x_6 + 3041\, x_7 \leq 372060$
2. $1000\, x_1 + 813\, x_2 + 1150\, x_3 + 805\, x_4 + 985\, x_5 + 1741\, x_6 + 870\, x_7 \geq 48300$
3. $x_1 + x_2 + x_3 + x_4 + x_5 + x_6 + x_7 + x_8 + x_9 + x_{10} + x_{11} = 60$
4. $x_1 \leq 15$
5. $x_2 \geq 8$
6. $x_3 \geq 15$
7. $x_4 \geq 10$
8. $x_5 \leq 8$
9. $x_6 \geq 5$
10. $x_7 \geq 5$

Compared with the previous option, the inferior limits of sugar beet and beans crops were changed from 4 ha to 5 ha, as they have a big profit per hectare. The other restrictions were still unchanged. As a result of applying the model, the following values for the options xj and for profit, gross margin and total expenses are obtained:

$$P_3 = \sum_{j=1}^{7} p_j \cdot x_j = 28186 \text{ lei};$$

$$MB_3 = \sum_{j=1}^{7} mj_j \cdot x_j = 61739 \text{ lei};$$

$$Ch_3 = \sum_{j=1}^{11} c_j \cdot x_j = 187264 \text{ lei}.$$

In this option, the following production structure is identified:

Wheat: $x_1 = 9$ ha;
Barley: $x_2 = 8$ ha;
Corn: $x_3 = 15$ ha;
Sunflower: $x_4 = 10$ ha;
Rapeseed: $x_5 = 8$ ha;
Sugar Beet: $x_6 = 5$ ha;
Beans: $x_7 = 5$ ha.

In the third option of the production structure, each 5 ha are allocated to sugar beet and beans crops, compared with 4 ha as they had in the second option, which does not lead to an increased profit, because the redistribution of the area was made on the basis of reducing the wheat crop, the most profitable of the crops in the production structure. In these conditions, we have tried a new option of optimization by introducing new restrictions regarding the limits of other crops.

Variant 4

In the fourth option, the target function is the same but a set of new restrictions is added to the model, which is:

$$\max(f(x)) = 607\,x_1 + 374\,x_2 + 399\,x_3 + 398\,x_4 + 647\,x_5 + 479\,x_6 + 439\,x_7$$

The restrictions imposed to the model below are:

1. $3073\,x_1 + 3026\,x_2 + 3161\,x_3 + 2231\,x_4 + 2433\,x_5 + 6201\,x_6 + 3041\,x_7 \leq 372060$
2. $1000\,x_1 + 813\,x_2 + 1150\,x_3 + 805\,x_4 + 985\,x_5 + 1741\,x_6 + 870\,x_7 \geq 48300$
3. $x_1 + x_2 + x_3 + x_4 + x_5 + x_6 + x_7 + x_8 + x_9 + x_{10} + x_{11} = 60$
4. $x_1 \leq 15$
5. $x_2 \geq 8$
6. $x_3 \geq 18$
7. $x_4 \geq 12$
8. $x_5 \leq 10$

9. $x_6 \geq 4$

10. $x_7 \geq 4$

Compared with the previous option, the inferior limits of the sugar beet and beans crops were changed from 5 ha to 4 ha and those of the corn crops from 15 ha to 18 ha and sunflower from 10 to 12 ha. As a result of running the plan, the following values for options x_j and for profit, gross margin and total expenses are obtained:

$$P_4 = \sum_{j=1}^{7} p_j \cdot x_j = 27520 \text{ lei};$$

$$MB_4 = \sum_{j=1}^{7} mj_j \cdot x_j = 61158 \text{ lei};$$

$$Ch_4 = \sum_{j=1}^{11} c_j \cdot x_j = 181468 \text{ lei}.$$

The following production structure is identified for this option of optimization:

Wheat: $x_1 = 4$ ha;
Barley: $x_2 = 8$ ha;
Corn: $x_3 = 18$ ha;
Sunflower: $x_4 = 12$ ha;
Rapeseed: $x_5 = 10$ ha;
Sugar Beet: $x_6 = 4$ ha;
Beans: $x_7 = 4$ ha.

In the fourth option, the production structure includes all crops, but profitability is reduced, because the area allocated to wheat, which is the crop having the biggest profit per hectare, reduced to 4 ha. That is why a new option with limits referring to the wheat crop cultivated on a larger area has been tried.

Variant 5

In the fifth option, even though the target function is still unchanged, the model is improved and new restrictions are applied as follows:

$$\max(f(x)) = 607\,x_1 + 374\,x_2 + 399\,x_3 + 398\,x_4 + 647\,x_5 + 479\,x_6 + 439\,x_7$$

The set of restrictions imposed to this option is:

1. $3073\ x_1 + 3026\ x_2 + 3161\ x_3 + 2231\ x_4 + 2433\ x_5 + 6201\ x_6 + 3041\ x_7 \leq 372060$
2. $1000\ x_1 + 813\ x_2 + 1150\ x_3 + 805\ x_4 + 985\ x_5 + 1741\ x_6 + 870\ x_7 \geq 48300$
3. $x_1 + x_2 + x_3 + x_4 + x_5 + x_6 + x_7 + x_8 + x_9 + x_{10} + x_{11} = 60$
4. $x_1 \leq 18$
5. $x_2 \geq 6$
6. $x_3 \geq 15$
7. $x_4 \geq 8$
8. $x_5 \leq 12$
9. $x_6 \geq 5$
10. $x_7 \geq 5$

Different from the fourth option, the superior limit for wheat was established at 18 ha, starting from the idea that due to running the plan, a larger area would be allocated to this crop, because on one hand it is fit for large areas and on the other hand it has the biggest profit per hectare, after rapeseed.

Barley is a less profitable crop that is why its limit was established to 6 ha, compared with corn, a crop having big profit per hectare, for which a limit of 15 ha was established. It was also imposed the condition that sunflower which is a product greatly demanded on market should be cultivated on an area of at least 8 ha.

Rapeseed, the most profitable crop has a limit of 12 ha (if it had not been introduced that condition it would have been allocated the largest area and the production structure would have been unbalanced). Sugar beet and beans crops have limits of at least 5 ha, being necessary in the production structure as they grant the rotation and minimize the risk associated with products selling. As a result of running the plan, the following values for the options xj and for profit, gross margin and total expenses are obtained:

$$P_5 = \sum_{j=1}^{7} p_j \cdot x_j = 29230 \text{ lei};$$

$$MB_5 = \sum_{j=1}^{7} mj_j \cdot x_j = 62443 \text{ lei};$$

$$Ch_5 = \sum_{j=1}^{11} c_j \cdot x_j = 186482 \text{ lei}.$$

As a result of applying the optimization method the following production structure is defined:

Wheat: $x_1 = 9$ ha;
Barley: $x_2 = 6$ ha;
Corn: $x_3 = 15$ ha;
Sunflower i: $x_4 = 8$ ha;
Rapeseed: $x_5 = 12$ ha;
Sugar Beet: $x_6 = 5$ ha;
Beans: $x_7 = 5$ ha.

In the fifth option, which is the last one realized within this production module, compared with the previous version, the area cultivated with wheat was enlarged from 4 ha to 9 ha, which means an additional profit, while the area cultivated with barley was reduced from 8 ha to 6 ha, the one cultivated with corn from 18 ha to 15 ha, and the one cultivated with sunflower from 12 ha to 8 ha.

The areas cultivated with rapeseed, sugar beet and beans, which are plants with big profit per hectare, were enlarged, which determined an increase of the profitability on the whole. In the table 3.2 the results of applying the linear plan method are centralized for the five options previously presented.

Table 3.2: Results of optimizing the production structure, per versions, for the 60 hectares module

Specification	UM	Option					
		0 (basic)	*1*	2	*3*	*4*	*5*
Wheat	ha	13	26	11	9	4	9
Barley	ha	7	8,5	8	8	8	6
Corn	ha	17	8,5	15	15	18	*15*
Sunflower	ha	10	8,5	10	10	12	*8*
Rapeseed	ha	6	8,5	8	8	10	*12*

Specification	UM	Option					
		0 (basic)	*1*	*2*	*3*	*4*	*5*
Sugar Beet	ha	3	0	4	5	4	*5*
Beans	ha	4	0	4	5	4	*5*
Total profit	lei	28347	31235	28482	28186	27520	*29230*
Gross margin	lei	60904	57900	61128	61739	61158	*62443*
Total expenses	lei	182543	172131	184168	187264	181468	*186482*
Profit rate	%	15,5	18	15,4	15	15,1	*15,6*

Source: author's own computation

With reference to maximizing the profit and gross margin, the optimal production structure is represented by the fifth option, which compared with the basic option (zero option) increases the profit from 28347 lei to 29230 lei (103%), the gross margin from 60904 lei to 62443 lei (102,5%), and the profit rate from 15,5% to 15,6%.

Compared with the basic option, the optimized one allocates a double area for rapeseed crop, approximately equal areas for barley, sunflower, sugar beet and beans, and a reduced area for wheat. As for minimizing the expenses, the cheapest option is fourth one, the expenses being reduced with 0.5% compared with the basic option and with 2.6% less than the option considered to be optimal regarding the profitability (option 5). It was not taken into account the first option, which does not offer a good rotation or an increased level of total gross margin, even though it records the highest level of total profit.

The optimal production structure (option 5) is viable not only economically, as it determines high levels of profit and gross margin, but also ecologically, as due to various crops such as: grains, oleaginous plants, leguminous and technical plants, rational rotation can be realized having positive results during the next years. The in-put data and the results of optimization calculation using the linear plan method are detailed in the Appendix 3.

The research can be continued by realizing future projects for the production structure using the mathematical model of the target functions aiming the maximization of the profit ($f(x)$), expenses ($g(x)$) and gross margin ($h(x)$), presented below:

$$\max(f(x)) = 607\,x_1 + 374\,x_2 + 399\,x_3 + 398\,x_4 + 647\,x_5 + 479\,x_6 + 439\,x_7$$

$$\min(g(x)) = 3073\,x_1 + 3026\,x_2 + 3161\,x_3 + 2231\,x_4 + 2433\,x_5 + 6201\,x_6 + 3041\,x_7$$

$$\max(h(x)) = 1000\,x_1 + 813\,x_2 + 1150\,x_3 + 805\,x_4 + 985\,x_5 + 1741\,x_6 + 870\,x_7$$

Variables $x_1 - x_7$ represent areas allocated to different crops and have values between 0 and 60. By choosing the values of variables, we can obtain levels of profit (function $f(x)$), expenses (function $g(x)$) and gross margin (function $h(x)$).

The prediction of the profit for the module of 60 ha during 2013–2017 was made firstly by predicting the prices of the agricultural products and production costs per hectare for which it was estimated an increase of 5% per year proportionally with the expected rate of inflation.

Replacing the price levels (price appendix) in the crops budgets for the 60 ha module and increasing the costs level with 5% per year we can obtain the following levels of the profit per year for wheat, barley, corn, sunflower, rapeseed, sugar beet and beans included in the farming plan of the module (table 3.3).

Table 3.3: Estimating the profit per hectare for crops in the 60 ha farm, 2013–2017

Crop	Profit (lei/ha)				
	2013	2014	2015	2016	2017
Wheat	663	720	775	861	947
Barley	383	456	498	541	615
Corn	433	499	534	568	667
Sunflower	417	449	471	505	554
Rapeseed	685	740	786	848	926
Sugar Beet	569	659	687	915	1143
Beans	479	535	576	634	707

Source: author's own computation

From the data presented (table 2.3 and diagram 2.1) we can notice that the level of profit per hectare for all crops in the production plan increases during 2013–2017 especially due to high levels of this indicator for the following crops: sugar beet, rapeseed and wheat.

In case of wheat crop we estimate an increase of the profit per hectare from 663 lei in 2013 to 947 lei in 2017, for rapeseed from 685 lei/ha in 2012 to 926 lei/ha in 2017 whereas sunflower records low levels from 417 lei/ha in 2012 to 554 lei/ha in 2017. In the fig 2.1 is presented the evolution of the profit per hectare for the farming having the area of 60 ha, for the year 2017.

Fig. 3.1: Trends of the profit/ha for 60 ha farms, during, 2013–2017 (lei/ha)

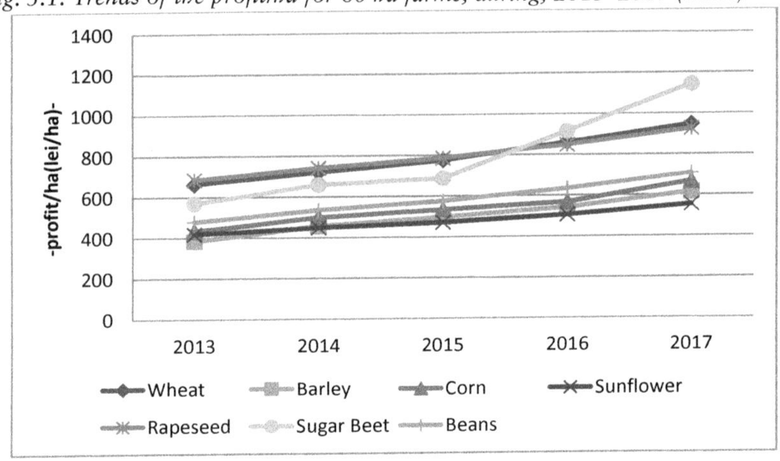

Source: based on data in Table 3.3

The total profit of the farming is estimated under these conditions, taking into account the optimal production structure, mentioned in table 3.2. The total profit follows the same trend as the unit one, especially due to high levels of profit generated by rapeseed, corn and wheat crops, cultivated on large areas. Table 3.4 and figure 3.2 present the predicted values of the total profit and its dynamics for the years 2013–2017.

Table 3.4: Predicting the total profit for the module of 60 ha, 2013–2017

Specification	Area (ha)	Total profit (lei)				
		2013	2014	2015	2016	2017
Wheat	9	5967	6480	6975	7749	8523
Barley	6	2298	2736	2988	3246	3690
Corn	15	6495	7485	8010	8520	10005

Specification	Area (ha)	Total profit (lei)				
		2013	2014	2015	2016	2017
Sunflower	8	3336	3592	3768	4040	4432
Rapeseed	12	8220	8880	9432	10176	11112
Sugar Beet	5	2845	3295	3435	4575	5715
Beans	5	2395	2675	2880	3170	3535
TOTAL	*60*	*31556*	*35143*	*37488*	*41476*	*47012*

Source: author's own computation

The fig. 3.2 presents the dynamics of the total profit of the agricultural farming having an area of 60 ha, for the year 2017, taking into account the calculation in the table 4.4. Obviously, the evolution of the total profit synchronizes with the evolution of the profit per hectare, recording a significant increase for the whole period.

Fig. 3.2: The evolution of the total profit estimated for the 60 ha farm (lei)

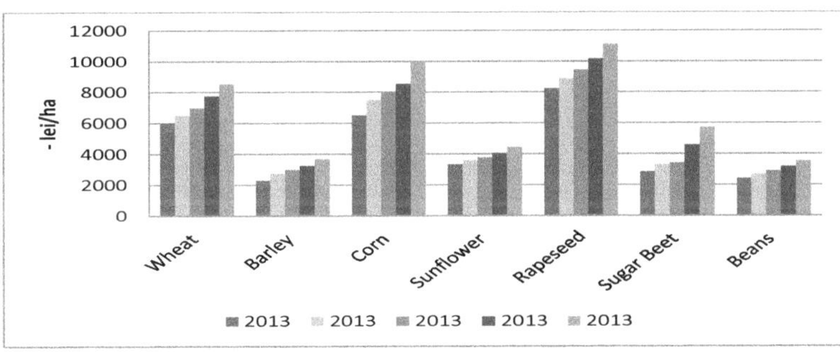

Source: based on data in Table 3.4

The results of the optimization of the production structure indicate the increase of the total profit, in case of reducing the areas cultivated with wheat, barley, corn and sunflower and enlarging the areas cultivated with rapeseed, sugar beet and beans. For the following period, the total profit of the farming increases, given that the costs per hectare increase slower than the sales prices of the agricultural products.

Starting from the idea that in practice cannot be used only the optimal crop but the conditions referring to the rotation and crops structure must be also observed, the crop budget for 2017 was set (appendix 8b), as well as the multiannual budget of the module of 60 ha, for the years 2013–2017, (appendix 8), taking into account the requirements previously mentioned, referring to the structure and rotation of crops. Table 3.6 presents the evolution of the main indicators corresponding to the module of 60 ha, for the years 2013–2017, taking into account these requirements.

Table 3.5: The main efficiency indicators for the module of 60 ha area

-lei/expl.-

No.	Indicator	2013	2014	2015	2016	2017
1	The value of main and secondary production	184662,0	197286,0	210590,0	225174,0	241360,0
2	Gross Product	225462,0	238086,0	251390,0	265974,0	282160,0
3	Total expenses	191661,0	200786,9	211741,5	222700,0	233643,9
4	Profit	33801,0	37299,1	39648,5	43274,0	48516,1
5	Profit rate (%)	16,0	17,0	17,2	18,0	19,4
6	Gross margin total farming	72269,5	84893,5	98197,5	112781,5	128967,5
7	Development fund (50% of profit)	16900,5	18649,6	19824,3	21637,0	24258,1
8	Funds for resuming the process of production (15% of profit)	5070,2	5594,9	5947,3	6491,1	7277,4
9	Funds for capitalization (25% of profit)	8450,2	9324,8	9912,1	10818,5	12129,0

Source: author's own computation based on data from Appendix 8

Based on the analysis of the indicators presented we can assert that by establishing a vegetal agricultural farming of 60 ha, in the field area and observing the conditions mentioned in this simulation, satisfactory economic results can be obtained if we take into account at least the evolution of the profit rate from 15.5% in 2012 to 19.4% in the year 2017.

The level of the economic results grants the funds necessary for resuming the process of production by allocating 155 of profit, increasing from 4737.4 lei/farming in 2012 to 7277.4 lei/farming in 2017. Although significant sums are allocated for the Fund of development (50% of profit) and the Funds for capitalization (25% of profit), it is necessary to increase the planned performance and reduce certain costs of production especially the material expenses and labor force.

3.2 Designing a farming module with 100 ha surface, in lowland agricultural area

A second solution tested in order to improve the usage of agricultural potential, is represented by the foundation and identification of an optimal crop production plan, for a 100 ha farm, established in Romanian plain agricultural area. Other recent researches have proposed agricultural crop production schemes for modules of 100 ha area, but they refer to different structures and opportunities as for the crop allocation. The present simulation is differentiating by the proposed crop rotation structure, the rotation scheme, as well as the number of simulated alternatives, in order to select the optimal production solution. The option for simulating a module of 100 ha had in mind that the 7992 existing holdings of 50–100 ha surface, use a total area of 563,500 ha (INS, 2012).

Design of the 100 ha module was performed maintaining the similar restrictions to the module of 60 ha, regarding crop rotation, rotation scheme and the crop production plan pack, the acreage being the only variable that has changed. In these circumstances, it was considered a five-year rotation with seven simple sole (Appendix 2), specific to the plain agricultural area, which included autumn grain (wheat + barley), hoes (corn + sunflower), crops (rapeseed + sugar beet) and legumes (beans).

The projected module of 100 ha is part of VII class size, with medium large economic size having a moderate economic viability (Appendix 7). This is corresponding to a profitability rate of 21.7%, provided to an average profit of 733.2 lei per ha for this type of establishment. Construction of basic (zero) version *for this agricultural production module is based on the initial values recorded* for expenditures, gross margin and profit per hectare per each crop in first crop rotation (Table 3.6).

Table 3.6: Production structure, costs, gross margin and profit per hectare in the basic version of the 100 ha farm

Crop	Area (ha)	Costs (lei/ha)	Gross margin (lei/ha)	Profit (lei/ha)
Wheat	22	3164	1124	716
Barley	11	3107	916	463
Maize	22	3206	1301	534
Sunflower	10	2280	911	500
Rape	12	2586	1152	794
Sugar beet	5	6330	2535	1250
Beans	12	3229	1412	951

Source: Author's own calculations based on the original budget

There might be observed that the record crops of sunflower and rapeseed entitles the lowest costs per hectare (2280 lei and 2586 lei); a gross margin of over 1000 lei per ha is provided by crops of sugar beet, beans, corn, oilseed rape and wheat, while high levels of yield of crops per hectare are generated by sugar beet, beans, rape and wheat (between 700 and 1,250 lei per ha). The initial production structure defines the following levels of total expenditure, gross margin and profit:

$$Ch_0 = \sum_{j=1}^{7} ch_j \cdot x_j = 298547 \text{ lei};$$

$$MB_0 = \sum_{j=1}^{7} mb_j \cdot x_j = 115979 \text{ lei};$$

$$P_0 = \sum_{j=1}^{7} p_j \cdot x_j = 64783 \text{ lei}.$$

where:

Ch – total expenditure;
MB – total gross margin;
P – total profit;
ch_j – production costs per unit for crop j;
mb_j – gross margin per unit of output for crop j;

p_j – profit per unit of output for the crop j;
x_j – area allocated for crop j.

Variant 1

The first production option is built using the objective function which aims to maximize the total profit of the production mode and the optimum choice of crops structure through the production program.

The considered model is:

$$\max(f(x)) = 716\ x_1 + 463\ x_2 + 534\ x_3 + 500\ x_4 + 794\ x_5 + 1250\ x_6 + 951\ x_7$$

Subject to the following set of restrictions:

1. $3164\ x_1 + 3107\ x_2 + 3206\ x_3 + 2280\ x_4 + 2586\ x_5 + 6330\ x_6 + 3229\ x_7 \leq 633000$
2. $1124\ x_1 + 916\ x_2 + 1301\ x_3 + 911\ x_4 + 1152\ x_5 + 2535\ x_6 + 1412\ x_7 \geq 91100$
3. $x_1 + x_2 + x_3 + x_4 + x_5 + x_6 + x_7 + x_8 + x_9 + x_{10} + x_{11} = 100$
4. $x_j \geq 0$

The first restriction of the model is in compliance to the maximum level of expenditure budget, taking into account costs per crop and the sum do not exceed a budget of 633000 lei.

The maximum limit of budget expenditures per crop was calculated by multiplying the maximum expenditure per hectare (in this Version, for sugar beet) by the total area of the module; this situation is afferent to the worst case scenario, according to that the entire area would be assigned to the most expensive crop.

The second restriction imposes a lower limit for the total gross margin of the module, which should not be less than 91,100 lei. The calculation of this value was done, as in the case of the 60 ha module, by multiplying the minimum gross margin per hectare (in this case, the crop of sunflower) and the total area. This reflects, as above, the worst-case scenario, when the entire surface would be allotted to the crop with lowest level of gross margin per hectare.

The third restriction imposes the harnessing of the entire surface of the module and the sum of the variables (in this case, the crops areas) does not exceed the total area of the module, that is, 100 hectares. As a result of applying the program, there are achieved the following values for profit, gross margin and total expenditures:

$$P_1 = \sum_{j=1}^{7} p_j \cdot x_j = 125000 \text{ lei};$$

$$MB_1 = \sum_{j=1}^{7} mj_j \cdot x_j = 253500 \text{ lei};$$

$$Ch_1 = \sum_{j=1}^{11} c_j \cdot x_j = 633000 \text{ lei}.$$

According to the following production structure:

Wheat: $x_1 = 0$ ha;
Barley: $x_2 = 0$ ha;
Corn: $x_3 = 0$ ha;
Sunflower: $x_4 = 0$ ha;
Rape: $x_5 = 0$ ha;
Sugar beet: $x_6 = 100$ ha;
Beans: $x_7 = 0$ ha.

In this version of the production structure, the whole area was allotted to sugar beet – the crop that ensures the highest level of profit per hectare. The solution is not viable in terms of crop rotation; therefore, there are required new simulations of production options, expressed by insertion of additional restrictions.

Variant 2

In this option, keeping the same objective function of optimization program, the constraints take supplementary into account while achieving crop rotation.

The model is:

$$\max(f(x)) = 716\, x_1 + 463\, x_2 + 534\, x_3 + 500\, x_4 + 794\, x_5 + 1250\, x_6 + 951\, x_7$$

Subject of restrictions:

1. $3164\,x_1 + 3107\,x_2 + 3206\,x_3 + 2280\,x_4 + 2586\,x_5 + 6330\,x_6 + 3229\,x_7 \leq 633000$
2. $1124\,x_1 + 916\,x_2 + 1301\,x_3 + 911\,x_4 + 1152\,x_5 + 2535\,x_6 + 1412\,x_7 \geq 91100$
3. $x_1 + x_2 + x_3 + x_4 + x_5 + x_6 + x_7 + x_8 + x_9 + x_{10} + x_{11} = 100$
4. $x_j \geq 0$
5. $x_1 \geq 20$
6. $x_2 \geq 10$
7. $x_3 \geq 20$
8. $x_4 \geq 10$
9. $x_5 \geq 10$
10. $x_6 \leq 10$

In order to introduce the cereals and oilseeds within the production structure, there were imposed lower limits: 20 ha for wheat, 10 ha for barley, 20 ha for maize, 10 ha of sunflower and 10 ha for rape. Sugar beet crop was limited to 10 hectares. After applying the optimization program, result the following values for profit, gross margin and total expenditures:

$$P_2 = \sum_{j=1}^{7} p_j \cdot x_j = 74090 \text{ lei};$$

$$MB_2 = \sum_{j=1}^{7} mj_j \cdot x_j = 131880 \text{ lei};$$

$$Ch_2 = \sum_{j=1}^{11} c_j \cdot x_j = 335010 \text{ lei}.$$

According to the following production structure:
Wheat: $x_1 = 20$ ha;
Barley: $x_2 = 10$ hectares;
Corn: $x_3 = 20$ ha;
Sunflower $x_4 = 10$ ha;

Rape: x_5 = 10 hectares;
Sugar beet: x_6 = 10 ha;
Beans: x_7 = 20 ha.

The Version 2 is in compliance to the conditions of crop rotation, as the production structure includes cereals (wheat, barley, maize), oilseeds (sunflower and rapeseed), technical crops (sugar beet) and legumes (beans – a good forerunner for cereals).

As the area allocated for beans crop is too big, this would hinder the carrying out works, and would imply additional labor costs. Therefore, in the new production option, new restrictions are imposed on variable x_7.

Variant 3

In this Version, the objective functions of optimization remains the same, but the restrictions take into account both rotations and their level of profitability. The model becomes:

$$\max(f(x)) = 716\, x_1 + 463\, x_2 + 534\, x_3 + 500\, x_4 + 794\, x_5 + 1250\, x_6 + 951\, x_7$$

Subject of restrictions:

1. $3164\, x_1 + 3107\, x_2 + 3206\, x_3 + 2280\, x_4 + 2586\, x_5 + 6330\, x_6 + 3229\, x_7 \leq 633000$
2. $1124\, x_1 + 916\, x_2 + 1301\, x_3 + 911\, x_4 + 1152\, x_5 + 2535\, x_6 + 1412\, x_7 \geq 91100$
3. $x_1 + x_2 + x_3 + x_4 + x_5 + x_6 + x_7 + x_8 + x_9 + x_{10} + x_{11} = 100$
4. $x_j \geq 0$
5. $x_1 \geq 20$
6. $x_2 \geq 10$
7. $x_3 \geq 20$
8. $x_4 \geq 10$
9. $x_5 \geq 10$
10. $x_6 \leq 10$
11. $x_7 \leq 10$

As a result of the program, there are obtained the following values for profit, gross margin and total expenditures:

$$P_3 = \sum_{j=1}^{7} p_j \cdot x_j = 72520 \text{ lei;}$$

$$MB_3 = \sum_{j=1}^{7} mj_j \cdot x_j = 129280 \text{ lei;}$$

$$Ch_3 = \sum_{j=1}^{11} c_j \cdot x_j = 328580 \text{ lei.}$$

According to the following production structure:

Wheat: x_1 = 20 ha;
Barley: x_2 = 10 hectares;
Corn: x_3 = 20 ha;
Sunflower: x_4 = 10 ha;
Rape: x_5 = 20 ha;
Sugar beet: x_6 = 10 ha;
Beans: x_7 = 10 ha.

In this version of the production structure, large areas are allocated for most profitable crops: cereals (wheat and maize) and rape (20 ha each). The remaining crops use 10 hectares each in total area.

Variant 4

In this Version, although the model objective function remains the same, it is improved by a set of further restrictions:

$$\max(f(x)) = 716x_1 + 463x_2 + 534x_3 + 500x_4 + 794x_5 + 1250x_6 + 951x_7$$

Restrictions:

1. $3164\,x_1 + 3107\,x_2 + 3206\,x_3 + 2280\,x_4 + 2586\,x_5 + 6330\,x_6 + 3229\,x_7 \leq 633000$
2. $1124\,x_1 + 916\,x_2 + 1301\,x_3 + 911\,x_4 + 1152\,x_5 + 2535\,x_6 + 1412\,x_7 \geq 91100$

3. $x_1 + x_2 + x_3 + x_4 + x_5 + x_6 + x_7 + x_8 + x_9 + x_{10} + x_{11} = 100$

4. $x_j \geq 0$

5. $x_1 \geq 20$

6. $x_2 \geq 10$

7. $x_3 \geq 20$

8. $x_4 \geq 15$

9. $x_5 \geq 12$

10. $x_6 \leq 8$

11. $x_7 \leq 15$

Compared to the previous version, there were modified lower limits for sunflower (from 10 to 15 ha) and rape (from 10 to 12 hectares) crops, as the acreages assigned to them in the third version was lower. In addition, there were changed the limits of variables x_6 and x_7 (sugar beet and beans), from 10 to 8 ha, and from 10 to 15 hectares, respectively. Optimization program provides the following values for profit, gross margin and total expenditures:

$$P_4 = \sum_{j=1}^{7} p_j \cdot x_j = 70923 \text{ lei};$$

$$MB_4 = \sum_{j=1}^{7} mj_j \cdot x_j = 126609 \text{ lei};$$

$$Ch_4 = \sum_{j=1}^{11} c_j \cdot x_j = 322777 \text{ lei}.$$

According to the following production structure:

Wheat: $x_1 = 20$ ha;
Barley: $x_2 = 10$ hectares;
Corn: $x_3 = 20$ ha;
Sunflower: $x_4 = 15$ ha;
Rape: $x_5 = 12$ ha;
Sugar beet: $x_6 = 8$ ha;
Beans: $x_7 = 15$ ha.

In this version, the production structure is balanced, as it includes all crops, but profitability is reduced, which implies testing of a new version more likely to generate a higher level of profit.

Variant 5

In this latter simulated Version, the objective function remains unchanged, but there are added new restrictions within the model:

$$\max(f(x)) = 716\,x_1 + 463\,x_2 + 534\,x_3 + 500\,x_4 + 794\,x_5 + 1250\,x_6 + 951\,x_7$$

Restrictions:

1. $3164\,x_1 + 3107\,x_2 + 3206\,x_3 + 2280\,x_4 + 2586\,x_5 + 6330\,x_6 + 3229\,x_7 \leq 633000$
2. $1124\,x_1 + 916\,x_2 + 1301\,x_3 + 911\,x_4 + 1152\,x_5 + 2535\,x_6 + 1412\,x_7 \geq 91100$
3. $x_1 + x_2 + x_3 + x_4 + x_5 + x_6 + x_7 + x_8 + x_9 + x_{10} + x_{11} = 100$
4. $x_j \geq 0$
5. $x_1 \geq 20$
6. $x_2 \geq 10$
7. $x_3 \geq 20$
8. $x_4 \geq 15$
9. $x_5 \geq 15$
10. $x_6 \leq 10$
11. $x_7 \leq 15$

Compared to the fourth version, there was modified the limit of colza crop from 12 ha to 15 ha, as it is a crop that is established on large areas. Also, there was a relaxed restriction for sugar beet, as it is a crop with high profitability. The program provides the following values for profit, gross margin and total expenditures:

$$P_5 = \sum_{j=1}^{7} p_j \cdot x_j = 71050 \text{ lei;}$$

$$MB_5 = \sum_{j=1}^{7} mj_j \cdot x_j = 128075 \text{ lei;}$$

$$Ch_5 = \sum_{j=1}^{11} c_j \cdot x_j = 327050 \text{ lei.}$$

With the following production structure:

Wheat: x_1 = 20 ha;
Barley: x_2 = 10 hectares;
Corn: x_3 = 20 ha;
Sunflower: x_4 = 15 ha;
Rape: x_5 = 15 ha;
Sugar beet: x_6 = 10 ha;
Beans: x_7 = 10 ha.

Compared to the precedent Version, the acreage for colza increased from 12 ha to 15 ha, for sugar beet from 8 ha to 10 ha, respectively, and the area allocated for beans crop decreased from 15 ha to 10 ha. Expansion of surfaces allocated for crops with high profitability (sugar beet and rapeseed), determined increase in overall profitability. In Table 3.7 there are summarized the results of the linear programming method in the five options above.

Table 3.7: Results of optimizing the production structure, per versions, for the 100 hectares module

	Measurement unit	**Version**					
		0 (basic)	1	2	3	4	5
Wheat	ha	22	0	20	**20**	20	20
Barley	"	11	0	10	**10**	10	10
Corn	"	22	0	20	**20**	20	20
Sunflower	"	10	0	10	**10**	15	15
Rape	"	12	0	10	**20**	12	15
Sugar beet	"	5	100	10	**10**	8	10
Beans	"	12	0	20	**10**	15	10
Total profit	lei	*64783*	*125000*	*74090*	***72520***	*70923*	*71050*
Gross margin	lei	*115979*	*253500*	*131880*	***129280***	*126609*	*128075*
Total expenditure	lei	*298547*	*633000*	*335010*	***328580***	*322777*	*327050*
Profit rate	%	*21,7*	*19,7*	*22,1*	**22**	*21,9*	*21,7*

Source: author's own calculations

From the point of view of maximizing profit and gross margin, the optimal structure of production is provided by version 2, which, compared to the baseline scenario, the profit increase from 64783 lei to 74090 lei (115%), the gross margin of 115979 lei from 131880 lei (114%) and the rate of return from 21.7% to 22.1%. Unlike the basic version, the optimized production structure allocates a double area for sugar beet, approximately equal acreages to wheat, barley, maize, sunflower, and increased surface for beans crop.

But the production structure in version 2 is not economically viable because it allocates too much space to beans crop (20 ha, i.e. 20%). For all these reasons, the version 3 can be considered to be optimal from the point of view of profitability and gross margin (which increased to 112%, or 113% compared to the base solution), and ensures the equilibrium that exists between crops of production structure.

From the point of view of minimizing costs, the cheapest option that may be applied is the fourth, in which the expenditure is 1.8% lower than the version considered optimal in terms of profitability (version 3).

The option 1 was not taken into consideration; although it would ensure the highest level of total profit, it does not provide a viable crop rotation as the production structure includes only sugar beet.

The optimal production structure (version 3) is not only economically viable, since it determines high levels of profit and gross margin, but is also sustainable, due to the existence of various categories of crops: cereals, oilseeds, crops and legumes, that ensure their rational rotation with beneficial implications on the results of subsequent years.

Versions 4 and 5 provide increased profits to the basic version and can be taken into account in the decision determining the production structure in the coming years. Considering the results of the optimization process of the production structure, there were identified program optimization objective functions applied to agricultural module area of 100 hectares, to maximize profit, ($f(x)$), minimizing costs ($g(x)$) and maximizing gross margin ($h(x)$) with the following forms:

$$\max(f(x)) = 716\, x_1 + 463\, x_2 + 534\, x_3 + 500\, x_4 + 794\, x_5 + 1250\, x_6 + 951\, x_7$$

$$\max(g(x)) = 3164\,x_1 + 3107\,x_2 + 3206\,x_3 + 2280\,x_4 + 2586\,x_5 + 6330\,x_6 + 3229\,x_7$$

$$\max(h(x)) = 1124\,x_1 + 916\,x_2 + 1301\,x_3 + 911\,x_4 + 1152\,x_5 + 2535\,x_6 + 1412\,x_7$$

As above, the variables $x_1 - x_7$ represent areas allocated to different crops and take values between 0 and 100. By choosing the value of variables result levels of profit (the function ($f(x)$), expenditures (function ($g(x)$) and gross margin (function ($h(x)$).

Analogous to the previous version, research can be continued in order to make projections of future production structure, by using mathematical modeling and redefinition of objective functions for profit maximization ($f(x)$), minimizing costs ($g(x)$) and maximizing gross margin ($h(x)$).

Similar to the case of 60 ha module, the model estimates the economic results, that is, yield per hectare and total profit for agricultural production module with area of 100 ha, for years 2013–2017. In this respect, the prices of agricultural products are forecasted through linear regression; also, is expected an increase of costs per hectare by 5% per year. (Appendix 7)

Following the same calculation methodology, by replacing the estimated prices and costs, we get the picture of future economic performance for wheat, barley, maize, sunflower, rapeseed, sugar beet sugar and beans crops (Table 3.8).

Table 3.8: Estimated profit per hectare for 100 ha module, 2013–2017

-(lei/ha)-

Crop	**Profit (lei/ha)**				
	2013	**2014**	**2015**	**2016**	**2017**
Wheat	782	848	914	1012	1110
Barley	478	561	612	664	749
Corn	578	655	701	747	860
Sunflower	526	566	598	643	702
Rape	845	914	975	1054	1151
Sugar beet	1393	1537	1617	1927	2237
Beans	1029	1128	1214	1320	1446

Source: author's own calculations

The evolution of profit per hectare for the projected production module of 100 ha area is shown in fig 3.3. There might be observed the increase of profit during 2013 – 2017, for all crops considered in the production plan. Crops of sugar beet, bean, rape and wheat record high levels of profitability.

Fig. 3.3: Trends profit per hectare, for the holding of 100 ha, 2013–2017 (lei/ha)

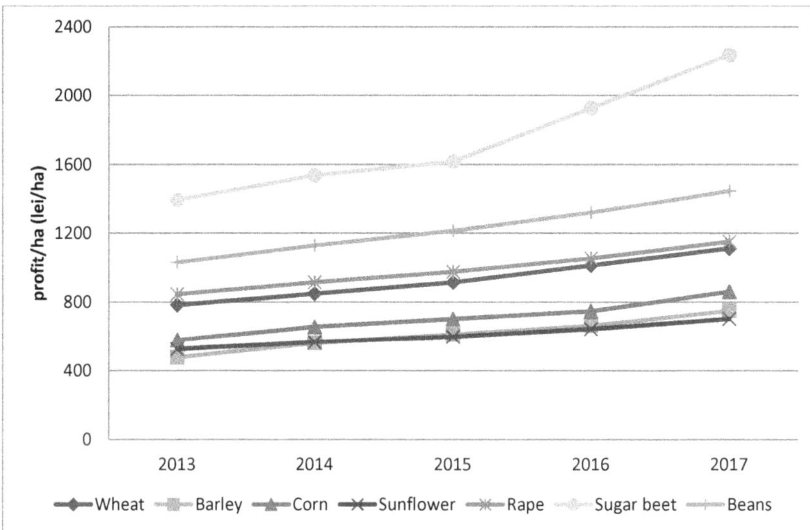

Source: based on data in Table 3.8

The previous data are used in calculation of the total profit for the 100 ha production module, by taking into account the optimal production structure. The provisions are shown the table 3.9.

Table 3.9: Projected total profit for the 100 ha production module

Crop	Area (ha)	Total profit (lei)				
		2013	2014	2015	2016	2017
Wheat	20	15640	16960	18280	20240	22200
Barley	10	4780	5610	6120	6640	7490
Corn	20	11560	13100	14020	14940	17200

Crop	Area (ha)	Total profit (lei)				
		2013	2014	2015	2016	2017
Sunflower	10	7890	8490	8970	9645	10530
Rape	20	12675	13710	14625	15810	17265
Sugar beet	10	13930	15370	16170	19270	22370
Beans	10	10290	11280	12140	13200	14460
TOTAL	*100*	*78360*	*86260*	*92210*	*101800*	*113760*

Source: author's own calculations

The total profit forecast for the agricultural module of 100 ha area confirms the natural growing trend (fig. 3.4).

Fig. 3.4: Trend of total profit forecast for the explorative agricultural production module of 100 ha area (lei)

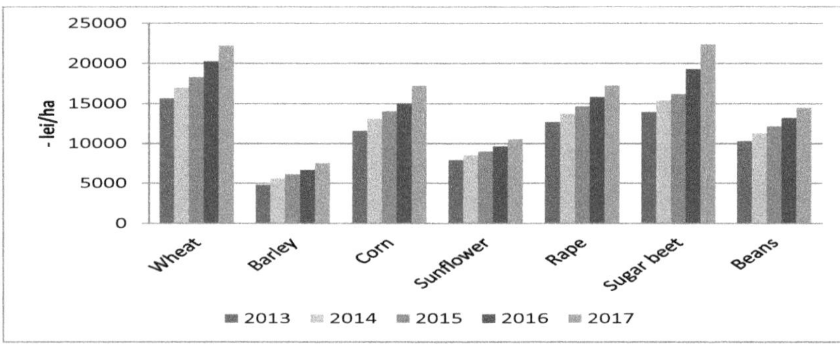

Source: based on data in Table 3.9

The results for the agricultural production module of 100 ha area express an increase in total profit, mainly based on high levels of profit for wheat and sugar beet crops – over 20000 lei. The results of the optimization in production structure show an increase in total profit if there are reduced wheat, barley, maize, and beans crops and grew those cultivated with rape and sugar beet. For the projected period, total profit of module increases, if the costs per hectare grow at a slower pace than sales prices of agricultural products.

In compliance with requirements of crop rotation and the crop structure in production planning are designed the multiannual budget (Appendix 9) and provisional budget of production module for the year 2017 (Appendix 9b).

The projected evolution of the main economic indicators for the production module of 100 ha area in 2013–2017 is described in Table 3.10.

Table 3.10: Summary of key efficiency indicators for 100 ha production module, 2013–2017

-lei/farm-

	Indicators	2013	2014	2015	2016	2017
1	Main and secondary production value	338409,0	361451,0	385877,0	412837,0	442347,0
2	Grants	68000,0	68000,0	68000,0	68000,0	68000,0
3	Gross product	406409,0	429451,0	453877,0	480837,0	510347,0
4	Total expenditures	327825,7	343437,1	362170,5	380903,9	399636,2
5	Profit	78583,3	86013,9	91706,5	99933,1	110710,8
6	Profitability (%)	22,3	23,4	23,8	24,8	26,3
7	Total Gross Margin	142302,8	165344,8	189770,8	216730,8	246240,8
8	Development Fund (50% of profits)	39291,7	43007,0	45853,3	49966,6	55355,4
9	Working capital (15% of profits)	11787,5	12902,1	13756,0	14990,0	16606,6
10	Capital funds (25% of profits)	19645,8	21503,5	22926,6	24983,3	27677,7

Source: Author's own calculations based on data from Appendix 19

Establishing of a 100 ha vegetable farm, in plain agricultural area, can be considered as a viable economic solution for harnessing a national agricultural potential. The evolution of profitability, from 21.7% in 2012 to 26.3% in 2017, confirms the economic viability of the farm.

However, the size of the development fund, established by allocating 50% of profits, create the possibility of investment in agricultural equipment (purchase of a tractor), but does not provide sufficient funds for sustainable development, which leads us to simulate to another farm module, of 250 hectares.

3.3 Designing a 250 hectare farming module, in lowland agricultural area

Simulation of farming production modules with area over 100 ha represents, for the current conditions of Romanian agriculture, an option for the future, while the farms with areas over 100 ha represent only 1.1% of the total farm in Romania, drawing 52.1% of eligible SPAS.

Under these conditions, we have designed a 250 ha agricultural module, in lowland agricultural area, which is a part of class VIII (Appendix 7.1), with an economic dimension considered as big, but with moderate economic viability, providing on average in the first working year, a profitability of 636.9 lei per ha, that is, a profit rate of 15.1%.

For this production module, it has been designed with a five-year rotation, with six simple soles (Appendix 3). Compared to previously simulated variants, there has been reduced by one the number of fields and the beans were replaced with soybean crop. Designing of Version 0 (basic) took into account the costs, gross margin and profit per hectare for each crop in the simulated production structure, considering the farm (module) budget in the first year of crop rotation (Appendix 10). The levels of these indicators are shown in Table 3.11.

Table 3.11: Production structure, expenditures, gross margin and profit per hectare in basic version, associated to the 250 ha module

Crop	Area (ha)	Expenditures (lei/ha)	Gross margin (lei/ha)	Profit (lei/ha)
Wheat	55	3278	1345	901,5
Barley	25	3188	1018	552
Corn	50	5335	1533	205,4
Sunflower	40	2207	1157	873
Rape	30	2736	1321	944
Soybeans	50	4791	976	89,5

Source: Author's own calculation based on the original budget

From calculations based on the original budget (Appendix 10), it can be observed that sunflower and rapeseed crops entitle the most reduced expenditures per hectare: 2207 lei and 2736 lei, respectively, while the gross margins for wheat, barley, corn, sunflower and rapeseed crops exceed 1000 lei/ha. The most elevated profits per hectare are those for rape, wheat and sunflower crops (944 lei/ha, 901 lei/ha, and 873 lei/ha, respectively). In exchange, corn and soybean generates smaller profits per hectare, but a significant contribution to environmental compliance requirements in the crop rotation. From the first observation, for this production module, the initial structure of production leads to the following figures of expenditure, gross margin and profit:

$$Ch_0 = \sum_{j=1}^{7} ch_j \cdot x_j = 936650 \text{ lei};$$

$$MB_0 = \sum_{j=1}^{7} mb_j \cdot x_j = 310785 \text{ lei};$$

$$P_0 = \sum_{j=1}^{7} p_j \cdot x_j = 141367{,}5 \text{ lei}.$$

Where:

Ch – total expenditure;
MB – total gross margin;
P – total profit;
ch_j – production costs per unit for crop *j*;
mb_j – gross margin per unit of output for crop *j*;
p_j – profit per unit of output for the crop *j*;
x_j – area allocated for crop *j*.

Variant 1

The first simulated version in optimization of production structure for 250 ha module, has as the objective function of the model, the optimization of crops structure by maximizing the total profit. The proposed model is:

$$\max (f(x)) = 901{,}5\, x_1 + 552\, x_2 + 205\, x_3 + 873\, x_4 + 944\, x_5 + 89{,}5\, x_6$$

And the restrictions:

1. $3278\, x_1 + 3188\, x_2 + 5335\, x_3 + 2207\, x_4 + 2736\, x_5 + 4791\, x_6 \leq 1333750$
2. $1345\, x_1 + 1018\, x_2 + 1533\, x_3 + 1157\, x_4 + 1321\, x_5 + 976\, x_6 \geq 244000$
3. $x_1 + x_2 + x_3 + x_4 + x_5 + x_6 + x_7 + x_8 + x_9 + x_{10} + x_{11} = 250$
4. $x_j \geq 0$

The set of restrictions imposed on this model envisages, analogous to the first two modules simulated, that the maximum level of expenditures for each crop and total expenditure not to exceed the total budget of 1333750 lei, resulted by multiplying the maximum expenditure per hectare (corresponding to the maize crop), with total area of 250 ha of the module. The minimum gross margin should be at least equal to 244000 lei (resulted from multiplying the minimum gross margin, in case of soybean crop, by the total area) starting from the same pessimistic premise that the entire area would be cultivated with the most expensive crop of the production plan. The model keeps the condition imposed by the third restriction: the sum of the variables (crop areas) should not exceed the aggregate total area of 250 ha. As result of applying the optimization program of the production structure, w have the following values for profit, gross margin and total expenditures:

$$P_1 = \sum_{j=1}^{7} p_j \cdot x_j = 236000 \text{ lei};$$

$$MB_1 = \sum_{j=1}^{7} mj_j \cdot x_j = 330250 \text{ lei};$$

$$Ch_1 = \sum_{j=1}^{11} c_j \cdot x_j = 684000 \text{ lei}.$$

based on the following production structure:

Wheat: $x_1 = 0$ ha;
Barley: $x_2 = 0$ ha;
Corn: $x_3 = 0$ ha;
Sunflower: $x_4 = 0$ ha;
Rapeseed: $x_5 = 250$ ha;
Soybeans: $x_7 = 0$ ha.

In this version of production structure, the whole area was allocated to rapeseed – the crop with highest profitability. The solution is not viable in terms of crop rotation to achieve therefore a new version, additional constraints are introduced.

Variant 2

In the second version, it is maintained the same objective function of the model, but restrictions take into account crop rotation, as follows:

$$\max(f(x)) = 901{,}5\ x_1 + 552\ x_2 + 205\ x_3 + 873\ x_4 + 944\ x_5 + 89{,}5\ x_6$$

And the restrictions:

1. $3278\ x_1 + 3188\ x_2 + 5335\ x_3 + 2207\ x_4 + 2736\ x_5 + 4791\ x_6 \leq 1333750$
2. $1345\ x_1 + 1018\ x_2 + 1533\ x_3 + 1157\ x_4 + 1321\ x_5 + 976\ x_6 \geq 244000$
3. $x_1 + x_2 + x_3 + x_4 + x_5 + x_6 + x_7 + x_8 + x_9 + x_{10} + x_{11} = 250$
4. $x_1 \geq 50$
5. $x_2 \geq 20$
6. $x_3 \geq 50$
7. $x_4 \geq 40$
8. $x_5 \geq 40$
9. $x_6 \geq 30$

To comply with restrictions imposed by crop rotation, that is, in order to introduce cereals and oilseeds within the production structure, there were imposed lower crop limits: 50 ha for wheat, 20 ha for barley, 50 ha for maize, 40 hectares for sunflower, 40 ha for rape and 30 ha in soybean. The result of the optimization program consists from the following values for profit, gross margin and total expenditures:

$$P_2 = \sum_{j=1}^{7} p_j \cdot x_j = 160610 \text{ lei};$$

$$MB_2 = \sum_{j=1}^{7} mj_j \cdot x_j = 319080 \text{ lei};$$

$$Ch_2 = \sum_{j=1}^{11} c_j \cdot x_j = 890580 \text{ lei}.$$

According to the following production structure:

Wheat: $x_1 = 20$ ha;
Barley: $x_2 = 20$ hectares;
Corn: $x_3 = 50$ ha;
Sunflower: $x_4 = 40$ ha;

Rape: $x_5 = 60$ ha;
Soybeans: $x_6 = 30$ ha.

The results of this version are in compliance with the conditions of crop rotation, as the structure of production includes both grains (wheat, barley, corn) and oilseeds (sunflower and rapeseed). Based on these findings, a new version comprises both restrictions regarding the variables and increases the total profit.

Variant 3

In this version, although the objective function remains unchanged, the restrictions take into account both the crop rotation, and their profitability. The model is:

$$\max(f(x)) = 901,5\ x_1 + 552\ x_2 + 205\ x_3 + 873\ x_4 + 944\ x_5 + 89,5\ x_6$$

And the restrictions:

1. $3278\ x_1 + 3188\ x_2 + 5335\ x_3 + 2207\ x_4 + 2736\ x_5 + 4791\ x_6 \leq 1333750$
2. $1345\ x_1 + 1018\ x_2 + 1533\ x_3 + 1157\ x_4 + 1321\ x_5 + 976\ x_6 \geq 244000$
3. $x_1 + x_2 + x_3 + x_4 + x_5 + x_6 + x_7 + x_8 + x_9 + x_{10} + x_{11} = 250$
4. $x_1 \geq 50$
5. $x_2 \geq 20$
6. $x_3 \geq 40$
7. $x_4 \geq 50$
8. $x_5 \geq 40$
9. $x_6 \geq 30$

As result of simulation of this version, there might be observed that the lower limit for maize crops (that induces economic losses) decreased from 50 ha to 40 ha. At the same time, the lower limit of sunflower, a high profitability crop increases from 40 to 50 hectares. The application of program yielded the following values for profit, gross margin and total expenditures:

$$P_3 = \sum_{j=1}^{7} p_j \cdot x_j = 167290 \text{ lei};$$

$$MB_3 = \sum_{j=1}^{7} mj_j \cdot x_j = 315320 \text{ lei};$$

$$Ch_3 = \sum_{j=1}^{11} c_j \cdot x_j = 859300 \text{ lei}.$$

Defining the following production structure:

Wheat: x_1 = 50 ha;
Barley: x_2 = 20 hectares;
Corn: x_3 = 40 ha;
Sunflower: x_4 = 50 ha;
Rape: x_5 = 60 ha;
Soybeans: x_6 = 30 ha.

In comparison to the previous version, there might be observed a decrease of area allocated to maize from 50 to 40 ha hectare, and an increase of sunflower acreage from 40 to 50 hectares.

Variant 4

In this version, the objective function remains unchanged, but the model is improved by new restrictions. The objective function:

$$\max(f(x)) = 901{,}5\ x_1 + 552\ x_2 + 205\ x_3 + 873\ x_4 + 944\ x_5 + 89{,}5\ x_6$$

Subject of the restrictions:

1. $3278\ x_1 + 3188\ x_2 + 5335\ x_3 + 2207\ x_4 + 2736\ x_5 + 4791\ x_6 \leq 1333750$
2. $1345\ x_1 + 1018\ x_2 + 1533\ x_3 + 1157\ x_4 + 1321\ x_5 + 976\ x_6 \geq 244000$
3. $x_1 + x_2 + x_3 + x_4 + x_5 + x_6 + x_7 + x_8 + x_9 + x_{10} + x_{11} = 250$
4. $x_1 \geq 50$
5. $x_2 \geq 20$
6. $x_3 \geq 40$
7. $x_4 \geq 50$
8. $x_5 \geq 40$
9. $x_6 \geq 30$

Compared to the previous version, there were changed the lower limit of the rapeseed crop, from 40 to 30 hectares, and introduced an upper limit of 50 for the soybean crop ha, since it is uneconomic. As a result of the program, the following values were obtained for profit, gross margin and total expenditures:

$$P_4 = \sum_{j=1}^{7} p_j \cdot x_j = 166865 \text{ lei};$$

$$MB_4 = \sum_{j=1}^{7} mj_j \cdot x_j = 315560 \text{ lei};$$

$$Ch_4 = \sum_{j=1}^{11} c_j \cdot x_j = 864720 \text{ lei}.$$

According to the following production structure:

Wheat: $x_1 = 50$ ha;
Barley: $x_2 = 20$ hectares;
Corn: $x_3 = 40$ ha;
Sunflower: $x_4 = 50$ ha;
Rape: $x_5 = 60$ ha;
Soybeans: $x_6 = 30$ ha.

In this production version, the wheat acreage increases from 50 to 60 ha, whereas the rape decreased from 60 ha to 50 ha. The production structure is balanced and includes all crops, but profitability is reduced; as result there is tried a new production version, that may entitles a higher level of profit.

Variant 5

Simulation is continued by creating a new version of the production structure, although it retains the same objective function, the new model is improved by adding new restrictions:

$$\max(f(x)) = 901,5\, x_1 + 552\, x_2 + 205\, x_3 + 873\, x_4 + 914\, x_5 + 89,5\, x_6$$

And the restrictions:

1. $3278\, x_1 + 3188\, x_2 + 5335\, x_3 + 2207\, x_4 + 2736\, x_5 + 4791\, x_6 \leq 1333750$

2. $1345\,x_1 + 1018\,x_2 + 1533\,x_3 + 1157\,x_4 + 1321\,x_5 + 976\,x_6 \geq 244000$
3. $x_1 + x_2 + x_3 + x_4 + x_5 + x_6 + x_7 + x_8 + x_9 + x_{10} + x_{11} = 250$
4. $x_1 \geq 50$
5. $x_2 \geq 20$
6. $x_3 \geq 35$
7. $x_4 \geq 50$
8. $x_5 \geq 25$
9. $x_6 \geq 30$

Compared to the fourth version, it was modified the maize crop limit from 40 to 35 ha, as it is a crop of loss. For the rape crop is specified a range between 40 and 50 ha and for soybean crop is imposed a lower acreage limit of at least 25 ha. After the application program yielded the following values for profit, gross margin and total expenditures:

$$P_5 = \sum_{j=1}^{7} p_j \cdot x_j = 174407 \text{ lei};$$

$$MB_5 = \sum_{j=1}^{7} mj_j \cdot x_j = 316465 \text{ lei};$$

$$Ch_5 = \sum_{j=1}^{11} c_j \cdot x_j = 846870 \text{ lei}.$$

With the following production structure:

Wheat: $x_1 = 70$ ha;
Barley: $x_2 = 20$ ha;
Maize: $x_3 = 35$ ha;
Sunflower: $x_4 = 50$ ha;
Rapeseed: $x_5 = 50$ ha;
Soybeans: $x_6 = 25$ ha.

In the version 5, compared to the previous one, the area of grain increases from 60 to 70 ha, the areas occupied by corn and soybean decrease from 40 to 35 hectares, and from 30 to 25 hectares, respectively. There has been increased the area allocated to wheat, a crop with high profitability, and reduced the areas allocated to maize and soya (crops of loss), thereby overall profitability increases. In the table 3.12 there are summarized the results of the linear programming method, for the 250 ha production module in the five options above.

Table 3.12: Results of optimizing the production structure, by versions, for the 250 ha module

	Measurement unit	Version					
		0 (basic)	1	2	3	4	5
Wheat	ha	55	0	50	50	60	70
Barley	"	25	0	20	20	20	20
Corn	"	50	0	50	40	40	35
Sunflower	"	40	0	40	50	50	50
Rape	"	30	250	60	60	50	50
Soybeans	"	50	0	30	30	30	25
Total profit	lei	*141367*	*236000*	*160610*	*167290*	*166865*	*174407*
Gross margin	lei	*310785*	*330250*	*319080*	*315320*	*315560*	*316465*
Total expenditure	lei	*936650*	*684000*	*890580*	*859300*	*864720*	*846870*
Profit rate	%	*15*	*34,5*	*18*	*19,4*	*19,3*	*20,6*

Source: author's own calculations

The optimal production structure is defined by the version 5, wherein, as compared to the basic version, the profit increase to 123%, the gross margin to 102%, and the rate of return from 15% to 20.6%. The optimized version allocates large areas for wheat, sunflower and rapeseed, crops with high returns per hectare and lower acreages for other crops.

From the point of view of minimizing costs, the fifth version is still the cheapest, entitling expenditures 9.6% lower than in the baseline scenario.

Also, there was not taken in account the first version: although it would imply the highest overall profit, it would not ensure a good crop rotation, as the structure containing only the rapeseed crop.

The optimal production structure (version 5) is viable not only economically, since it ensures the highest levels of profit and gross margin, but is also sustainable due to the existence of various categories of crops (cereals, oilseeds and legumes) that can ensure their rational rotation, with beneficial implications on the results of subsequent years.

It is possible to forecast the future production structure using mathematical modeling of objective functions for profit maximization ($f(x)$), costs minimization ($g(x)$), and maximization of gross margin ($h(x)$), as follows:

$$\max(f(x)) = 901,5\ x_1 + 552\ x_2 + 205\ x_3 + 873\ x_4 + 944\ x_5 + 89,5\ x_6$$

$$\min(g(x)) = 3278\ x_1 + 3188\ x_2 + 5335\ x_3 + 2207\ x_4 + 2736\ x_5 + 4791\ x_6$$

$$\max(h(x)) = 1345\ x_1 + 1018\ x_2 + 1533\ x_3 + 1157\ x_4 + 1321\ x_5 + 976\ x_6$$

The variables x_1 ,..., x_6 identify the areas allocated to different crops and take values between 0 and 250. Through the choice of values and their replacement achieved the levels of profit (the function ($f(x)$), expenditures (function ($g(x)$) and gross margin (function ($g(x)$).

Analogous to other production modules, there may be projected economic results for the period 2013–2017, by applying same procedures; first, estimating prices of agricultural products by linear regression (see Appendix 7) and the costs per hectare by applying an increase of 5% per year.

Using price levels (Appendix 7) in crop budgets for 250 hectares module and increasing the cost by 5% per year, the estimated economic results for wheat, barley, maize, sunflower, rapeseed, and soybeans crops are presented in the table 3.13.

Table 3.13: Estimated profit per hectare for crops of 250 ha module, 2013–2017 (lei/ ha)

Crop	Total profit (lei)				
	2013	2014	2015	2016	2017
Wheat	983	1064	1147	1265	1384
Barley	573	665	726	787	883
Corn	850	989	1081	1172	1372
Sunflower	922	988	1048	1123	1215
Rapeseed	1007	1090	1166	1262	1378
Soybeans	619	723	804	904	1025

Source author's own calculations

In Figure 3.5 there are shown the trends in profits per hectare for crops included in the production plan of the farm of 250 ha, between 2013 to 2017 forecasts, based on the results in Table 2.13.

Fig. 3.5: Trends in profit per hectare for farm of 250 ha, 2013–2017 (lei/ ha)

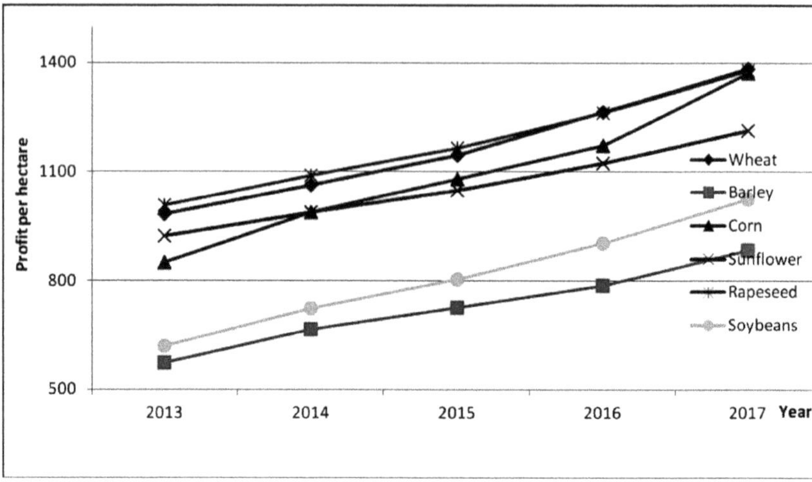

Source: based on data in Table 3.13

Under these conditions the profit increases for the whole period 2013–2017, for all cultures. Colza, wheat, maize and sunflower entitle important profitability. The simulation of total profit level is performed considering the optimal structure of production, that is, version 5.

Table 3.14: Projection of total profit for 250 ha production module, 2013–2017

Crop	Area (ha)	Total profit (lei)				
		2013	2014	2015	2016	2017
Wheat	70	68810	74480	80290	88550	96880
Barley	20	11460	13300	14520	15740	17660
Corn	35	29750	34615	37835	41020	48020
Sunflower	50	46100	49400	52400	56150	60750
Rape	50	50350	54500	58300	63100	68900
Soybeans	25	15475	18075	20100	22600	25625
TOTAL	*250*	*221945*	*244370*	*263445*	*287160*	*317835*

Source: author's own calculations

From the data of table 3.14, there can be observed that the total profit increased for all crops in the production plan. High levels of profit are projected for wheat, corn, sunflower and rapeseed, as large areas where these cultures are set up. In graph 2.6 there is presented the evolution of total gross margin of the area of 250 ha farm, in the perspective of 2017.

Fig. 3.6: The forecast for the total profit trends for the 250 ha production module, 2013–2017 lei)

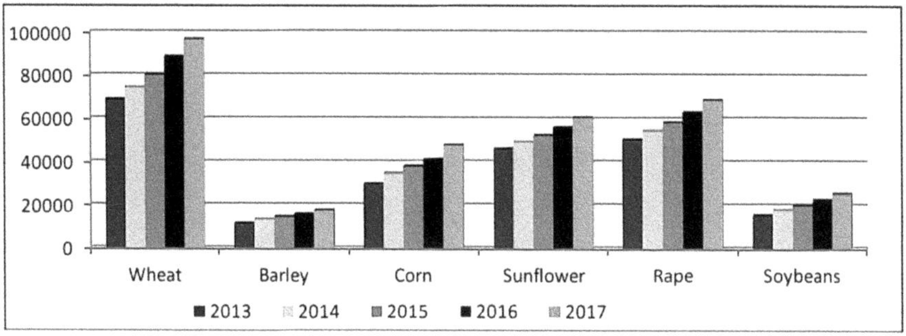

Source: based on the results in Table 3.14

As a conclusion, the estimation of total profit in five years was based on optimizing the production structure, by using linear programming, through forecasting agricultural prices (using linear regression) and costs (using the average growth rate method).

The results of production structure optimization express an increase in total profit as there are reduced acreages allocated to crops of barley, maize and soybean, on the one hand, and growing acreages allocated to crops of with wheat, sunflower and rapeseed. For the next period, the total profit of farm increases whilst the costs per hectare grow at a slower pace than sales prices of agricultural products.

As previously, in the case of this production module were drawn up projections of the multiannual budget (see Appendix 10) and of the annual budget for each crop at the end of the interval, for 2017, respectively (Appendix 10b), in compliance with the requirements of crop rotation and production structure. The projection of key economic indicators related to the 250 ha production module, in anticipation of 2017, is presented in Table 3.15.

Table 3.15: Summary of key efficiency indicators for 250 ha production module, 2013–2017

-lei/farm-

No.	Indicators	2013	2014	2015	2016	2017
1	The value of main and secondary production	986465,0	1054780,0	1127260,0	1203905,0	1289090,0
2	Grants	170000,0	170000,0	170000,0	170000,0	170000,0
3	Gross product (1+2)	1156465,0	1224780,0	1297260,0	1373905,0	1459090,0
4	Total expenditures	929641,0	973910,5	1027028,0	1080150,0	1133273,0
5	Profit (3–4)	226824,0	250869,5	270232,0	293755,0	325817,0
6	Profitability (%)	22,5	23,9	24,6	25,5	27,2
7	Total Gross Margin	418572,3	486887,3	559367,3	636012,3	721197,3
8	Development Fund (50% of profits)	113412,0	125434,8	135116,0	146877,5	162908,5
9	Working capital (15% of profits)	34023,6	37630,4	40534,8	44063,3	48872,6
10	Capital funds (25% of profits)	56706,0	62717,4	67558,0	73438,8	81454,3

Source: Author's own calculations based on data from Appendix 20

Analyzing the figures in the table 3.15, there can be observed that both production value and the gross profit increase over the entire period, while the level of subsidies remains constant. Thus, if in 2012, this production module generates a profit rate of 21.8%, five years later it reached 27.2%. Both the rate of profit and the level of capital funds confirm the viability of establishing a production mode of this size. The amount of development fund creates the premises of investment in the purchase of a tractor and of basic agricultural machinery, but the funds generated are not enough for a wider development. Thus, we continued the research in agricultural production module by simulating a double agricultural acreage.

3.4 Forecasting a module of farming to an area of 500 ha, in lowland agricultural area

Testing the possibilities to exploit agricultural potential by designing agricultural production module, we have decided to simulate production of a module with an area of 500 ha, lowland area, while maintaining the same crop rotation and crop scheme and doubling the area from previous farm (Appendix 4).

Along with the boost in agricultural area, class size and yield economic holding increase. Designed module of 500 ha belongs to the ninth class of large economic size and high economic viability (Appendix 7.1). The agricultural module of 500 ha shows a rate of return of 20.8%, and provides an average profit of 882.9 lei/ha in the first year of operation. (Appendix 11a)

The Variant 0 (basic) of the model of production starts, as in the case of other modules projected, from the initial level of expenses, gross margin and profit per hectare (Table 3.16), for each crop within the initial plan of production that will be optimized (Appendix 4).

Table 3.16: Structure of production, expenses, gross margin and profit per hectare, for module 500 ha

Crop	Area (ha)	Expenses (lei/ha)	Gross margin (lei/ha)	Profit (lei/ha)
Wheat	100	3375	1543	1105
Barley	50	3347	1227	733
Corn	100	5455	1999	635
Sunflower	80	2447	1368	933
Rape	70	2880	1497	1100
Soybean	100	4893	1310	407

Source: own calculation based on initial budget

Therefore, considering the data in the table above, a low level of expenses per hectare has been registered in the case of sunflower and rape: 2447 lei and, respectively, 2880 lei. Corn, wheat, sunflower, and rape have high gross margins. High profits per hectare have been registered in the case of rape and wheat (1100–1105 lei/ha), while corn, barley and sunflower have low levels of profits per hectare.

Using the data in the table, the following structure of production is designed, with expenses, gross margin and profit:

$$Ch_0 = \sum_{j=1}^{7} ch_j \cdot x_j = 1937010 \text{ lei};$$

$$MB_0 = \sum_{j=1}^{7} mb_j \cdot x_j = 760780 \text{ lei};$$

$$P_0 = \sum_{j=1}^{7} p_j \cdot x_j = 402990 \text{ lei}.$$

Where:

Ch – total expenses;

MB – total gross margin;

P – total profit;

ch_j – expenses per unit of production for crop *j*;

mb_j – gross margin per unit of production for crop *j*;

p_j – profit per unit of production for crop *j*;

x_j – area of crop *j*.

Variant 1

The objective function of the first variant obtained by applying the method of linear programming aims to optimize the structure of crops by maximizing the total profit. The model is:

$$\max(f(x)) = 1105\,x_1 + 733\,x_2 + 625\,x_3 + 933\,x_4 + 1100\,x_5 + 407\,x_6$$

Restrictions:

1. $3375\,x_1 + 3347\,x_2 + 5455\,x_3 + 2447\,x_4 + 2880\,x_5 + 4893\,x_6 \leq 2727500$
2. $1543\,x_1 + 1227\,x_2 + 1999\,x_3 + 1368\,x_4 + 1497\,x_5 + 1310\,x_6 \geq 613500$

3. $x_1 + x_2 + x_3 + x_4 + x_5 + x_6 + x_7 + x_8 + x_9 + x_{10} + x_{11} = 500$

4. $x_j \geq 0$

The first restriction refers to the level of expenses for each crop; their sum is below the maximum budget of 2727500 lei. This value corresponds to the result of multiplying the maximum costs per hectare, in this case of the corn, with total area of 500 ha, starting from the pessimist premise that the entire area was cultivated with the most expensive culture.

The second restriction refers to the total gross margin which must be higher, at least equal, with 613500 lei. This value corresponds to the result of multiplying the minimum gross margin per hectare, in this case of the barley, with total area, starting from the premise that, in a pessimist scenario the entire surface would be cultivated with the crop with the lowest gross margin per hectare.

The third restriction refers to the sum of variables (crops' areas) that must not exceed the total area of 500 ha.

Applying the model of optimization, the following values for profit, gross margin and expenses are obtained:

$$P_1 = \sum_{j=1}^{7} p_j \cdot x_j = 552500 \text{ lei},$$

$$MB_1 = \sum_{j=1}^{7} mj_j \cdot x_j = 771500 \text{ lei},$$

$$Ch_1 = \sum_{j=1}^{11} c_j \cdot x_j = 1687500 \text{ lei}.$$

Corresponding to the following structure of production:

Wheat: $x_1 = 500$ ha;
Barley: $x_2 = 0$ ha;
Corn: $x_3 = 0$ ha;
Sunflower: $x_4 = 0$ ha;
Rape: $x_5 = 0$ ha;
Soybean: $x_6 = 0$ ha.

In this variant of structure of production, the entire area has been allocated to wheat – the crop with the highest level of profit per hectare. The solution is not valid as regards the crops rotation. A new variant, with supplementary restrictions, is tried.

Variant 2

In variant 2, the objective function is the same, but restrictions refer to the criteria related to crops rotation. The model is:

$$\max(f(x)) = 1105\, x_1 + 733\, x_2 + 625\, x_3 + 933\, x_4 + 1100\, x_5 + 407\, x_6$$

Restrictions:

1. $3375\, x_1 + 3347\, x_2 + 5455\, x_3 + 2447\, x_4 + 2880\, x_5 + 4893\, x_6 \leq 2727500$
2. $1543\, x_1 + 1227\, x_2 + 1999\, x_3 + 1368\, x_4 + 1497\, x_5 + 1310\, x_6 \geq 613500$
3. $x_1 + x_2 + x_3 + x_4 + x_5 + x_6 + x_7 + x_8 + x_9 + x_{10} + x_{11} = 500$
4. $x_1 \geq 80$
5. $x_2 \geq 80$
6. $x_3 \geq 80$
7. $x_4 \geq 80$
8. $x_5 \geq 80$
9. $x_6 \geq 80$

Lower limits for all crops have been established to 80 hectares, starting from the intention to grow equals areas. After running the program, the profit, gross margin and expenses are obtained:

$$P_2 = \sum_{j=1}^{7} p_j \cdot x_j = 415140 \text{ lei;}$$

$$MB_2 = \sum_{j=1}^{7} mj_j \cdot x_j = 746380 \text{ lei;}$$

$$Ch_2 = \sum_{j=1}^{11} c_j \cdot x_j = 1859260 \text{ lei.}$$

With the following structure of production:

Wheat: $x_1 = 100$ ha;
Barley: $x_2 = 80$ ha;
Corn: $x_3 = 80$ ha;
Sunflower: $x_4 = 80$ ha;
Rape: $x_5 = 80$ ha;
Soybean: $x_6 = 80$ ha.

Results obtained in Variant 2 respect the conditions regarding the crops' rotation, because the structure of production includes cereals (wheat, barley and corn), oilseeds crops (sunflower and rape) and legumes (soybean).

Another variant is tried with new restrictions related to variables.

Variant 3

In variant 3, the objective function is the same, but restrictions refer both to crops' rotation and their efficiency. The model is:

$$\max(f(x)) = 1105\,x_1 + 733\,x_2 + 625\,x_3 + 933\,x_4 + 1100\,x_5 + 407\,x_6$$

Restrictions:

1. $3375\,x_1 + 3347\,x_2 + 5455\,x_3 + 2447\,x_4 + 2880\,x_5 + 4893\,x_6 \le 2727500$
2. $1543\,x_1 + 1227\,x_2 + 1999\,x_3 + 1368\,x_4 + 1497\,x_5 + 1310\,x_6 \ge 613500$
3. $x_1 + x_2 + x_3 + x_4 + x_5 + x_6 + x_7 + x_8 + x_9 + x_{10} + x_{11} = 500$
4. $x_1 \ge 80$
5. $x_2 \ge 70$
6. $x_3 \ge 80$
7. $x_4 \ge 80$
8. $x_5 \ge 70$
9. $x_6 \ge 80$

Lower limits for barley and rape are implemented, because these crops usually use small areas within the structure of production. After running the program of optimization, the profit, gross margin and total expenses are as follows:

$$P_3 = \sum_{j=1}^{7} p_j \cdot x_j = 418910 \text{ lei};$$

$$MB_3 = \sum_{j=1}^{7} mj_j \cdot x_j = 750000 \text{ lei};$$

$$Ch_3 = \sum_{j=1}^{11} c_j \cdot x_j = 1864490 \text{ lei}.$$

With the following structure of production:

Wheat: $x_1 = 120$ ha;
Barley: $x_2 = 70$ ha;
Corn: $x_3 = 80$ ha;
Sunflower: $x_4 = 80$ ha;
Rape: $x_5 = 70$ ha.
Soybean: $x_6 = 80$ ha

Compared to Variant 2, in this areas cultivated with barley and rape decreased from 80 hectares to 70 hectares and the area cultivated with wheat increased from 100 hectares to 120 hectares.

Variant 4

In the Variant 4, the model of optimization is improved using new restrictions, under the same objective function. The model is:

$$\max(f(x)) = 1105\, x_1 + 733\, x_2 + 625\, x_3 + 933\, x_4 + 1100\, x_5 + 407\, x_6$$

Restrictions:

1. $3375\, x_1 + 3347\, x_2 + 5455\, x_3 + 2447\, x_4 + 2880\, x_5 + 4893\, x_6 \leq 2727500$
2. $1543\, x_1 + 1227\, x_2 + 1999\, x_3 + 1368\, x_4 + 1497\, x_5 + 1310\, x_6 \geq 613500$
3. $x_1 + x_2 + x_3 + x_4 + x_5 + x_6 + x_7 + x_8 + x_9 + x_{10} + x_{11} = 500$
4. $x_1 \geq 80$
5. $x_2 \geq 50$
6. $x_2 \leq 70$

7. $x_3 \geq 100$
8. $x_4 \geq 100$
9. $x_5 \geq 100$
10. $x_6 \geq 70$

Compared to the previous variant, lower and upper limits for variable x_2 have been implemented. They can take values between 50 and 70 hectares. Low limits to corn, sunflower and rape increased; corn and sunflower are usually cultivated on large areas and rape is a crop with high yields. Profit, gross margin and total expenses are:

$$P_4 = \sum_{j=1}^{7} p_j \cdot x_j = 420340 \text{ lei;}$$

$$MB_4 = \sum_{j=1}^{7} mj_j \cdot x_j = 762890 \text{ lei;}$$

$$Ch_4 = \sum_{j=1}^{11} c_j \cdot x_j = 1858060 \text{ lei.}$$

The structure of production is:

Wheat: $x_1 = 80$ ha;
Barley: $x_2 = 50$ ha;
Corn: $x_3 = 100$ ha;
Sunflower: $x_4 = 100$ ha;
Rape: $x_5 = 100$ ha;
Soybean: $x_6 = 70$ ha.

In Variant 4, the area cultivated with wheat has been reduced. This implies a lower level of profit. The areas with sunflower, corn and rape have been increased. The structure of production is balanced and contains all crops, but profitability is lower.

Variant 5
In Variant 5, the model of optimization is improved with new restrictions. The model is:

$$\max(f(x)) = 1105\,x_1 + 733\,x_2 + 625\,x_3 + 933\,x_4 + 1100\,x_5 + 407\,x_6$$

Restrictions:

1. $3375\,x_1 + 3347\,x_2 + 5455\,x_3 + 2447\,x_4 + 2880\,x_5 + 4893\,x_6 \leq 2727500$
2. $1543\,x_1 + 1227\,x_2 + 1999\,x_3 + 1368\,x_4 + 1497\,x_5 + 1310\,x_6 \geq 613500$
3. $x_1 + x_2 + x_3 + x_4 + x_5 + x_6 + x_7 + x_8 + x_9 + x_{10} + x_{11} = 500$
4. $x_1 \geq 100$
5. $x_2 \geq 50$
6. $x_2 \leq 70$
7. $x_3 \geq 80$
8. $x_4 \geq 100$
9. $x_5 \geq 100$
10. $x_6 \geq 70$

Compared to Variant 4, the limit of corn has been modified from 100 to 80 hectares, because it is a less efficient crop. Also, the lower limit of wheat increased from 80 to 100 hectares. After running the program, the profit, gross margin and total expenses are:

$$P_5 = \sum_{j=1}^{7} p_j \cdot x_j = 429740 \text{ lei;}$$

$$MB_5 = \sum_{j=1}^{7} mj_j \cdot x_j = 753770 \text{ lei;}$$

$$Ch_5 = \sum_{j=1}^{11} c_j \cdot x_j = 1816460 \text{ lei.}$$

With the structure of production:

Wheat: $x_1 = 100$ ha;
Barley: $x_2 = 50$ ha;
Corn: $x_3 = 80$ ha;
Sunflower: $x_4 = 100$ ha;
Rape: $x_5 = 100$ ha;
Soybean: $x_6 = 70$ ha.

In Variant 5, compared to the previous one, the area cultivated with wheat increased from 80 to 100 hectares and the area cultivated with corn decreased from 100 to 80 hectares. Increasing the area with wheat, which has high profit per hectare, and decreasing the area with corn, which has losses per hectare, lead to increasing the overall profitability. The results of the five variants are presented in Table 3.17.

Table 3.17: Results of optimizing the structure of production, with variants, for module of 500 hectares

Specification	**MU**	**Variant**					
		0	1	2	3	4	5
Wheat	ha	100	500	100	120	80	*100*
Barley	"	50	0	80	70	50	*50*
Corn	"	100	0	80	80	100	*80*
Sunflower	"	80	0	80	80	100	*100*
Rape	"	70	0	80	70	100	*100*
Soybean	"	100	0	80	80	70	*70*
Total profit	lei	*402990*	*552500*	*415140*	*418910*	*420340*	*429740*
Gross margin	lei	*760780*	*771500*	*746380*	*750000*	*762890*	*753770*
Total expenses	lei	*1937010*	*1687500*	*1859260*	*1864490*	*1858060*	*1816460*
Rate of return	%	*20,8*	*32,7*	*22,3*	*22,4*	*22,6*	*23,6*

Source: own calculations

The optimal structure of production is Variant 5 in which, compared to Variant 0, the profit increased to 106% and the rate of return rose from 20.8% to 23.6%. The optimized variant brings in larger areas cultivated with sunflower and rape, crops which have high profit per hectare, and lower areas for the other crops.

As regards the expenses, Variant 5 is still the cheapest one; the expenses are 6% lower compared to Variant 0. Variant 1 has not been taken into consideration, although it ensures the highest total profit, because it does not provide crops' rotation (it contains only one culture: wheat).

The optimal structure of production, Variant 5 is viable not only from economic point of view, because it ensures high levels of profit and gross margin, but from ecological point of view, because it provides all types of crops: cereals, oilseeds and legumes. Rational rotations can be made, with positive implications in the following years.

As in other modules, forecasts for future structures of production can be realized, using mathematical modeling of objective functions of maximizing the profit ($f(x)$), minimizing the expenses ($g(x)$) and minimizing the gross margin ($h(x)$):

$$\max(f(x)) = 1105\, x_1 + 733\, x_2 + 625\, x_3 + 933\, x_4 + 1100\, x_5 + 407\, x_6$$

$$\min(g(x)) = 3375\, x_1 + 3347\, x_2 + 5455\, x_3 + 2447\, x_4 + 2880\, x_5 + 4893\, x_6$$

$$\max(h(x)) = 1543\, x_1 + 1227\, x_2 + 1999\, x_3 + 1368\, x_4 + 1497\, x_5 + 1310\, x_6$$

Variables $x_1 - x_6$ represent the areas cultivated with different crops and take values between 0 and 500 hectares.

Economic results are forecasted for this module, in the period 2013–2017, by estimating the prices of agricultural products (Appendix 7), and the costs per hectare which will increase by 5% per year.

Filling in the level of prices in crops budgets for module of 500 hectares, and increasing the level of costs by 5% per year, the economic results for wheat, barley, corn, sunflower, rape and soybean are as follows (Table 3.18).

Table 3.18: Estimation of profit per hectare for crops in module of 500 hectares, 2013–2017

Crop	Profit (lei/ha)				
	2013	2014	2015	2016	2017
Wheat	1202	1300	1401	1541	1680
Barley	766	878	926	1037	1156
Corn	1299	1474	1601	1727	1973
Sunflower	991	1067	1136	1223	1328
Rape	1177	1275	1366	1479	1614
Soybean	960	1090	1199	1329	1481

Source: own calculations

The dynamic of profit per hectare, for module of 500 hectares, in the period 2013–2017, is illustrated in the graph bellow.

Fig. 3.7: Dynamic of profit per hectare, for module of 500 ha, 2013–2017 (lei/ha)

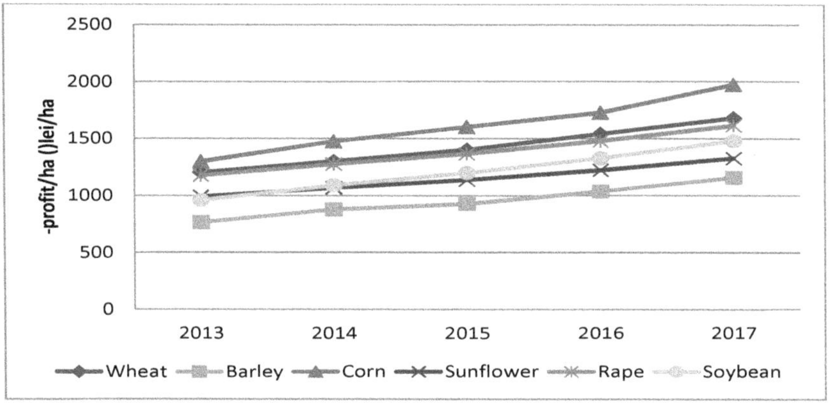

Source: data of table 2.18

The profit increases, in the period 2013–2017, for all products (wheat, rape, corn, soybean have the highest profits per hectare). The total profit is estimated considering the optimal structure of production, expressed in Table 3.18. The results of forecasting the total profit of the farm are presented in Table 3.19.

Table 3.19: Forecasting the total profit for module of 500 hectares, 2013–2017

Specification	Area (ha)	Total profit (lei)				
		2013	2014	2015	2016	2017
Wheat	100	120200	130000	140100	154100	168000
Barley	50	38300	43900	46300	51850	57800
Corn	80	103920	117920	128080	138160	157840
Sunflower	100	99100	106700	113600	122300	132800
Rape	100	117700	127500	136600	147900	161400
Soybean	70	67200	76300	83930	93030	103670
TOTAL	*500*	*546420*	*602320*	*648610*	*707340*	*781510*

Source: own calculations

In the period 2013–2017, the total profit increases for all crops within the plan of production. High levels of profit are registered for rape, corn and wheat, because they are cultivated on large areas. Under these conditions, the profit resulted from the 100 hectares of wheat, in 2013, is 120200 lei, for rape 117700 lei and the 80 hectares of corn produce a profit of 103900 lei. In fig. 3.8, the dynamic of total profit is represented, in view of 2017, for module of 500 hectares.

Fig. 3.8: Dynamic of total profit forecasted, for module of 500 ha, 2013–2017 (lei)

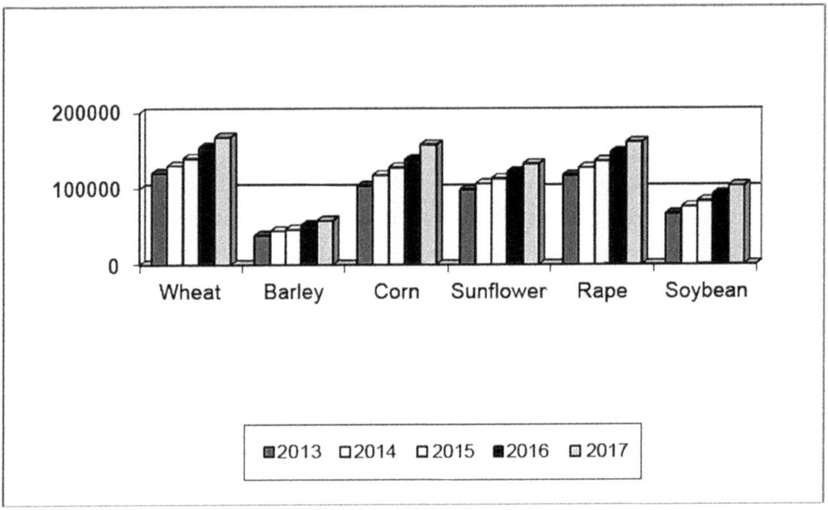

Source: based on data in Table 3.19

The results of program of optimization the structure of production, for module of 500 hectares, show an increase of total profit, if the areas with corn and soybean decrease and the areas with sunflower and rape increase. For the following period, the total profit increases to 143% from 2013 to 2017.

The multiannual budget for the module of 500 hectares (Appendix 11) and the annual budget of crops (Appendix 11b) are designed, for the year 2017. In the table below, the main economic indicators of the module of 500 hectares are presented, in view of 2013–2017, respecting the crops' rotation principle.

Table 3.20: Summary of main indicators of efficiency for module of 500 ha -

lei/expl-

No.	Indicators	2012	2013	2014	2015	2016	2017
1	Value of primary and secondary production	2038500.0	2171700.0	2322280.0	2481840.0	2650380.0	2838100.0
2	Subventions	340000.0	340000.0	340000.0	340000.0	340000.0	340000.0
3	Gross product (1+2)	2378500.0	2511700.0	2662280.0	2821840.0	2990380.0	3178100.0
4	Total expenses	1834454.6	1926163.0	2017887.0	2129512.0	2238010.0	2348080.0
5	Profit (3–4)	544045.4	585537.0	644393.0	692328.0	752370.0	830020.0
6	Rata of return (%)	27.5	28.4	30.0	30.7	31.9	33.7
7	Gross margin for total exploitation	855069.0	988269.0	1138849.0	1298409.0	1466949.0	1654669.0
9	Development fund (50% of profit)	272022.7	292768.5	322196.5	346164.0	376185.0	415010.0
10	Funds for resumption of production (15% of profit)	81606.8	87830.6	96659.0	103849.2	112855.5	124503.0
11	Funds for capitalization (25% of profit)	136011.4	146384.3	161098.3	173082.0	188092.5	207505.0

Source: own calculations based on data from Appendix 11

The data in the table above, compared with results from previous simulations, it can be argued that the establishment of agricultural production modules of 500 ha area enables the optimum conditions for recovery of the national agricultural potential, especially since development fund constituted creates the possibility of investment in agricultural equipment (purchase of tractors, combine harvesters and agricultural machinery base).

High levels of rate of return of 33.7% in the perspective of 2017, confirm the high performance for this mode of production, which can be an attractive investment, given increasing competitiveness of Romanian agriculture.

3.5 Forecasting a module of farming to an area of 800 ha, in lowland agricultural area

The module of 800 hectares belongs to the Class IX, with very large economic dimension (Appendix 6.1), and high economic viability. The rate of return is 24.5% and the profit is 1041.5 lei/hectare.

Determination of Variant 0 (the basic) has been achieved considering the level of expenses, gross margin and profit per hectare, for each crop, designing the farm's budget for the first year of exploitation (Table 3.21) and optimizing the structure of production (Appendix 5).

Table 3.21: Structure of production, expenses, gross margin and profit per hectare, for module 800 ha

Crop	Area (ha)	Expenses (lei/ha)	Gross margin (lei/ha)	Profit (lei/ha)
Wheat	160	3375	1543	1105
Barley	80	3347	1227	733
Corn	160	5576	2464	1044
Sunflower	128	2447	1368	933
Rape	112	2880	1497	1100
Soybean	160	4959	1677	761

Source: own calculations based on initial budget

Analysis of data in table above reveal low level of expenses per hectare for sunflower and rape: 2447 lei, respectively 2880 lei, and high level of gross margin for wheat.

High levels of profits are obtained to wheat, rape and corn ((1105 lei/ha, 1100 lei/ha, respectively 1044 lei/ha), while barley and soybean registered low levels of profit per hectare. This structure of production enables the following values for total expenses, gross margin and profit:

$$Ch_0 = \sum_{j=1}^{7} ch_j \cdot x_j = 3129136 \text{ lei};$$

$$MB_0 = \sum_{j=1}^{7} mb_j \cdot x_j = 1350368 \text{ lei};$$

$$P_0 = \sum_{j=1}^{7} p_j \cdot x_j = 766864 \text{ lei}.$$

Where:
Ch – total expenses;
MB – total gross margin;
P – total profit;
ch_j – expenses per unit of production for crop *j;*
mb_j – gross margin per unit of production for crop *j;*
p_j – profit per unit of production for crop *j;*
x_j – area of crop *j*.

Variant 1

To design the Variant 1 of the plan of production, structure optimization is taken into consideration, and total profit maximization for the whole farm. The model proposed has the form:

$$\max(f(x)) = 1105\, x_1 + 733\, x_2 + 1044\, x_3 + 933\, x_4 + 1100\, x_5 + 761\, x_6$$

Restrictions:

1. $3375\, x_1 + 3347\, x_2 + 5576\, x_3 + 2447\, x_4 + 2880\, x_5 + 4959\, x_6 \leq 4460800$
2. $1543\, x_1 + 1227\, x_2 + 2464\, x_3 + 1368\, x_4 + 1497\, x_5 + 1677\, x_6 \geq 981600$
3. $x_1 + x_2 + x_3 + x_4 + x_5 + x_6 + x_7 + x_8 + x_9 + x_{10} + x_{11} = 800$
4. $x_j \geq 0$

The first restriction refers to the level of expenses for each crop; their sum must be under the budget of 4460800 lei. This value has been obtained by multiplying the maximum expenses per hectare (corresponding to corn culture) and total area of 800 hectares, starting from the premise that, in a pessimistic scenario, all area would be cultivated with the most expensive crop.

The second restriction considers the total gross margin which must be higher, at least equal, to 981600 lei. This value has been obtained by multiplying the minimum gross margin per hectare (corresponding to barley culture) and total area, starting from the premise that, in a pessimistic scenario, all area would be cultivated with the crop that has the lowest level of gross margin per hectare.

The third restriction refers to the fact that the sum of variables (crops' areas) must not exceed the total area of 800 hectares. After running the program, the values for xj variables and profit, gross margin and total expenses are:

$$P_1 = \sum_{j=1}^{7} p_j \cdot x_j = 884000 \text{ lei;}$$

$$MB_1 = \sum_{j=1}^{7} mj_j \cdot x_j = 1234400 \text{ lei;}$$

$$Ch_1 = \sum_{j=1}^{11} c_j \cdot x_j = 2700000 \text{ lei.}$$

With the following structure of production:

Wheat: $x_1 = 800$ ha;
Barley: $x_2 = 0$ ha;
Corn: $x_3 = 0$ ha;
Sunflower: $x_4 = 0$ ha;
Rape: $x_5 = 0$ ha;
Soybean: $x_6 = 0$ ha.

In this variant, all area has been allocated to wheat – the crop with the highest level of profit per hectare. The solution is not viable from the point of view of crops' rotation, thus, a new variant is running, with supplementary restrictions.

Variant 2

In Variant 2, the objective function is the same, but restrictions take into consideration the crops' rotation as well. The simulated model is:

$$\max(f(x)) = 1105\,x_1 + 733\,x_2 + 1044\,x_3 + 933\,x_4 + 1100\,x_5 + 761\,x_6$$

Restrictions:

1. $3375\,x_1 + 3347\,x_2 + 5576\,x_3 + 2447\,x_4 + 2880\,x_5 + 4959\,x_6 \leq 4460800$
2. $1543\,x_1 + 1227\,x_2 + 2464\,x_3 + 1368\,x_4 + 1497\,x_5 + 1677\,x_6 \geq 981600$
3. $x_1 + x_2 + x_3 + x_4 + x_5 + x_6 + x_7 + x_8 + x_9 + x_{10} + x_{11} = 800$
4. $x_1 \geq 130$
5. $x_2 \geq 130$
6. $x_3 \geq 130$
7. $x_4 \geq 130$
8. $x_5 \geq 130$
9. $x_6 \geq 130$

In this variant, new restrictions have been introduced for all variables, such that crops have areas approximately equal. Running the program leads to the following values for variables x_j, profit, gross margin and total expenses:

$$P_2 = \sum_{j=1}^{7} p_j \cdot x_j = 759980 \text{ lei;}$$

$$MB_2 = \sum_{j=1}^{7} mj_j \cdot x_j = 1301740 \text{ lei;}$$

$$Ch_2 = \sum_{j=1}^{11} c_j \cdot x_j = 3003420 \text{ lei.}$$

With the following structure of production:

Wheat: $x_1 = 150$ ha;
Barley: $x_2 = 130$ ha;
Corn: $x_3 = 130$ ha;

Sunflower: $x_4 = 130$ ha;
Rape: $x_5 = 130$ ha;
Soybean: $x_6 = 130$ ha.

In Variant 2, crops' rotation principles are respected, because the structure of production contains all types of products: cereals (wheat, barley and corn), oilseeds (sunflower and rape) and legumes (soybean). In this situation, the crops have areas approximately equal.

Variant 3

In Variant 3, the objective function is the same, but restrictions take into consideration both crops' rotation and efficiency. The model is:

$$\max(f(x)) = 1105\,x_1 + 733\,x_2 + 1044\,x_3 + 933\,x_4 + 1100\,x_5 + 761\,x_6$$

Restrictions:

1. $3375\,x_1 + 3347\,x_2 + 5576\,x_3 + 2447\,x_4 + 2880\,x_5 + 4959\,x_6 \leq 4460800$
2. $1543\,x_1 + 1227\,x_2 + 2464\,x_3 + 1368\,x_4 + 1497\,x_5 + 1677\,x_6 \geq 981600$
3. $x_1 + x_2 + x_3 + x_4 + x_5 + x_6 + x_7 + x_8 + x_9 + x_{10} + x_{11} = 800$
4. $x_1 \geq 130$
5. $x_2 \geq 80$
6. $x_2 \leq 100$
7. $x_3 \geq 130$
8. $x_4 \geq 130$
9. $x_5 \geq 100$
10. $x_5 \leq 120$
11. $x_6 \geq 130$

Barley has an area between 80 and 100 hectares, because it returns a low level of profit per hectare and its cultivation on a large scale would mean lower total profit. Rape has an area between 100 and 120 hectares, because it a crop which uses small areas. The other restrictions remain unchanged. After running the program, the profit, gross margin and total expenses are as follows:

$$P_3 = \sum_{j=1}^{7} p_j \cdot x_j = 778730 \text{ lei};$$

$$MB_3 = \sum_{j=1}^{7} mj_j \cdot x_j = 1318920 \text{ lei;}$$

$$Ch_3 = \sum_{j=1}^{11} c_j \cdot x_j = 3019670 \text{ lei.}$$

With the structure of production:

Wheat: $x_1 = 230$ ha;
Barley: $x_2 = 80$ ha;
Corn: $x_3 = 130$ ha;
Sunflower: $x_4 = 130$ ha;
Rape: $x_5 = 100$ ha;
Soybean: $x_6 = 130$ ha.

Compared to the previous variant, the third one brings smaller areas cultivated with barley and rape. After running the program, the area cultivated with wheat increased – 230 hectares, which leads to an unbalanced structure of production. Thus, the optimization process continues with new restrictions.

Variant 4

This variant improves the results of the previous variant, under conditions of achieving the same objective. The model adds new restrictions:

$$\max(f(x)) = 1105\, x_1 + 733\, x_2 + 1044\, x_3 + 933\, x_4 + 1100\, x_5 + 761\, x_6$$

Restrictions:

1. $3375\, x_1 + 3347\, x_2 + 5576\, x_3 + 2447\, x_4 + 2880\, x_5 + 4959\, x_6 \leq 4460800$
2. $1543\, x_1 + 1227\, x_2 + 2464\, x_3 + 1368\, x_4 + 1497\, x_5 + 1677\, x_6 \geq 981600$
3. $x_1 + x_2 + x_3 + x_4 + x_5 + x_6 + x_7 + x_8 + x_9 + x_{10} + x_{11} = 800$
4. $x_1 \geq 130$
5. $x_1 \leq 200$
6. $x_2 \geq 80$
7. $x_2 \leq 100$
8. $x_3 \geq 130$

9. $x_4 \geq 130$

10. $x_5 \geq 100$

11. $x_5 \leq 120$

12. $x_6 \geq 130$

Compared to the previous variant, an upper limit of 200 hectares for wheat has been introduced, to balance the structure of production. After running the program, the values of profit, gross margin and total expenses are:

$$P_4 = \sum_{j=1}^{7} p_j \cdot x_j = 778020 \text{ lei};$$

$$MB_4 = \sum_{j=1}^{7} mj_j \cdot x_j = 1327210 \text{ lei};$$

$$Ch_4 = \sum_{j=1}^{11} c_j \cdot x_j = 3031780 \text{ lei}.$$

With the following structure of production:

Wheat: $x_1 = 200$ ha;
Barley: $x_2 = 80$ ha;
Corn: $x_3 = 140$ ha;
Sunflower: $x_4 = 130$ ha;
Rape: $x_5 = 120$ ha;
Soybean: $x_6 = 130$ ha.

In Variant 4, the area cultivated with wheat decreased from 230 hectares to 200 hectares; the difference has been allocated to corn and rape. The structure of production is balanced and contains all crops, but the efficiency is low. A new variant is tried, for increasing the profit.

Variant 5

In Variant 5, with the same objective function of maximizing the profit, the model continues to be improved by adding new restrictions.

The model of optimization is:

$$\max(f(x)) = 1105\, x_1 + 733\, x_2 + 1044\, x_3 + 933\, x_4 + 1100\, x_5 + 761\, x_6$$

With the following restrictions:

1. $3375\,x_1 + 3347\,x_2 + 5576\,x_3 + 2447\,x_4 + 2880\,x_5 + 4959\,x_6 \leq 4460800$
2. $1543\,x_1 + 1227\,x_2 + 2464\,x_3 + 1368\,x_4 + 1497\,x_5 + 1677\,x_6 \geq 981600$
3. $x_1 + x_2 + x_3 + x_4 + x_5 + x_6 + x_7 + x_8 + x_9 + x_{10} + x_{11} = 800$
4. $x_1 \geq 130$
5. $x_1 \leq 180$
6. $x_2 \geq 80$
7. $x_2 \leq 100$
8. $x_3 \geq 150$
9. $x_4 \geq 130$
10. $x_5 \geq 100$
11. $x_5 \leq 120$
12. $x_6 \geq 130$

Compared to Variant 4, the upper limit of wheat has been reduced from 200 hectares to 180 hectares, the lower limit of corn (high efficiency crop) increased from 130 hectares to 150 hectares. The values of profit, gross margin and total expenses are:

$$P_5 = \sum_{j=1}^{7} p_j \cdot x_j = 776800 \text{ lei;}$$

$$MB_5 = \sum_{j=1}^{7} mj_j \cdot x_j = 1345630 \text{ lei;}$$

$$Ch_5 = \sum_{j=1}^{11} c_j \cdot x_j = 3075800 \text{ lei;}$$

With the following structure of production:

Wheat: x_1 = 180 ha;
Barley: x_2 = 80 ha;
Corn: x_3 = 160 ha;
Sunflower: x_4 = 130 ha;

Rape: x_5 = 120 ha;
Soybean: x_6 = 130 ha.

In Variant 5, compared to the previous one, the area cultivated with wheat decreased from 200 hectares to 180 hectares, but it still holds the highest share within the structure of production. The area cultivated with corn increase from 140 hectares to 160 hectares. In Table 2.22, the results of applying the method of linear programming, with the five variants presented before, are centralized.

Table 3.22: Results of optimizing the structure of production, with variants, for module of 800 hectares

Specification	MU	Variant					
		0	1	2	3	4	5
Wheat	ha	160	800	150	230	*200*	180
Barley	"	80	0	130	80	*80*	80
Corn	"	160	0	130	130	*140*	160
Sunflower	"	128	0	130	130	*130*	130
Rape	"	112	0	130	100	*120*	120
Soybean	"	160	0	130	130	*130*	130
Total profit	lei	*766864*	*884000*	*759980*	*778730*	*778020*	*776800*
Gross margin	lei	*1350368*	*1234400*	*1301740*	*1318920*	*1327210*	*1345630*
Total expenses	lei	*3129136*	*2700000*	*3003420*	*3019670*	*3031780*	*3075800*
Rate of return	%	*24,5*	*32,7*	*25,3*	*25,8*	*25,7*	*25,3*

Source: own calculations

Variants 3, 4 and 5 are very closed in those regarding the values of profit and gross margin. Still, we consider that Variant 4 defines the optimal structure of production, because not only efficiency increases (compared to the basic variant, profit increased to 101.5% and the rate of return from 24.5% to 25.7%), but it is balanced considering the crops' rotation. Marginal profit increases as larger areas are allocated to wheat, but this cannot be done under disequilibrium among areas assigned to one crop or another.

As regards the expenses minimization, the cheapest variant is the one with appreciatively equal areas (Variant 2). Variant 1 has not been taken into consideration, although it brings the highest profit and the lower expenses, because it does not respect the principles of crops' rotation (it holds only one crop: wheat).

The optimal structure of production (Variant 4) is viable not only from the economic point of view, because enables achieving high levels of profit and gross margin, but also from ecological point of view, because it ensures the rational rotation of crops, with positive results in the following years.

Future projections of structure of production are realized, using mathematical modelling of functions of profit maximization ($f(x)$), expenses minimization ($g(x)$), and, respectively gross margin maximization ($h(x)$):

$$\max(f(x)) = 1105\,x_1 + 733\,x_2 + 1044\,x_3 + 933\,x_4 + 1100\,x_5 + 761\,x_6$$

$$\min(g(x)) = 3375\,x_1 + 3347\,x_2 + 5576\,x_3 + 2447\,x_4 + 2880\,x_5 + 4959\,x_6$$

$$\max(h(x)) = 1543\,x_1 + 1227\,x_2 + 2464\,x_3 + 1368\,x_4 + 1497\,x_5 + 1677\,x_6$$

Variables $x_1 - x_6$ represent the areas assigned to different crops and take values between 0 and 800 hectares. Forecasting economic results, for the period 2013–2017, has been realized as in the case of the other four modules of production, starting from the projection of products' prices (Appendix 7) and costs per hectare, which are estimated to increase by 5% per year.

Replacing the levels of prices in the budgets of crops for module of 800 hectares and increasing the level of costs by 5% per year, the economic results for wheat, barley, corn, sunflower, rape and soybean are as follows (Table 3.23).

Table 3.23: Estimation of profit per hectare for crops in module of 800 hectares, 2013–2017

Crop	Profit (lei/ha)				
	2013	2014	2015	2016	2017
Wheat	1202	1300	1401	1541	1680
Barley	766	878	958	1037	1156
Corn	1748	1959	2120	2281	2574
Sunflower	991	1067	1136	1223	1328
Rape	1177	1275	1366	1479	1614
Soybean	1338	1498	1636	1798	1985

Source: own calculations

Fig. 3.9: Dynamic of profit per hectare, for farm of 800 ha, 2013–2017 (lei/ha)

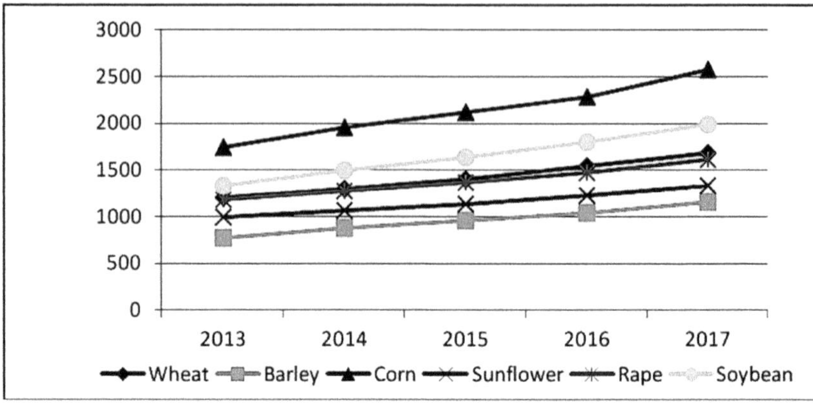

Source: based on data in Table 3.23

An increase in the level of profit per hectare for all crops can be noticed, for the entire period taken into consideration, which springs the idea of starting a business with this structure of production.

High levels of profit per hectare are evident for rape, from 1177 lei/ha in 2013 to 1614 lei/ha in 2017, wheat, from 1202 lei/ha in 2013 to 1680 lei/ha in 2017 and barley from 766 lei/ha in 2013 to 1165 lei/ha in 2017. In Graph 4.9, the evolution of profit per hectare is presented, in the period 2013–2017, for all crops included in program of production of the farm of 800 hectares.

The profit increases in the period 2013–2017 for all crops. Corn, wheat, rape and soybean have high profit per hectare. Total profit has been estimated considering the optimal structure of production (Variant 4). In Table 3.24 the total profit for all crops of the 800 hectares module is presented, for the period 2013–2017.

Table 3.24: Forecasting total profit for module of 800 ha, 2013–2017

Crop	Area (ha)	Total profit (lei)				
		2013	2014	2015	2016	2017
Wheat	200	240400	260000	280200	308200	336000
Barley	80	61280	70240	76640	82960	92480

Crop	Area (ha)	Total profit (lei)				
		2013	2014	2015	2016	2017
Corn	140	244720	274260	296800	319340	360360
Sunflower	130	128830	138710	147680	158990	172640
Rape	120	141240	153000	163920	177480	193680
Soybean	130	173940	194740	212680	233740	258050
TOTAL	800	990410	1090950	1177920	1280710	1413210

Source: own calculations

Total profit increases in the period 2013–2017, on account of high profit of wheat, corn, soybean and rape. Thus, in 2013, the 200 hectares cultivated with wheat generate a profit of 240400 lei, corn, cultivated on 140 hectare, returns a profit of 244700 lei and the 133 hectares cultivated with soybean generate a profit of 173900 lei. In fig 3.10, the evolution of total profit of the farm is presented, for the period 2013–2017.

Fig. 3.10: Dynamic of total profit forecasted, for module of 800 ha, 2013–2017 (lei)

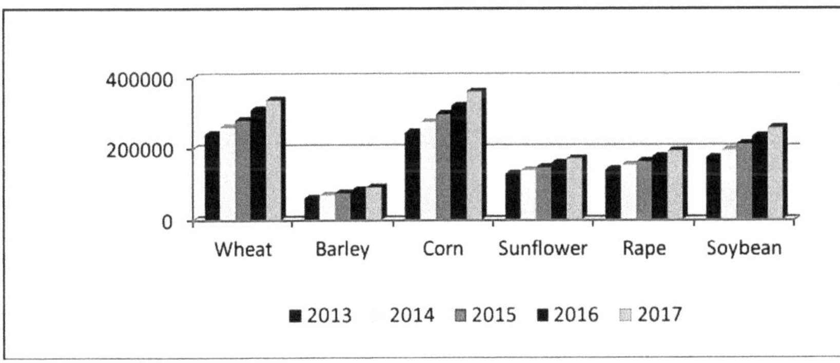

Source: based on data in Table 3.24

Applying the method of linear programming for optimizing the structure of production of the module of 800 hectares, leads to achieving superior economic results (Appendix 12), through increasing the total profit of exploitation, if the area cultivated with soybean has been reduced and areas assigned to wheat and rape increased. The total profit increases in period

2013–2017, to 142%. In Table 3.25 the evolution of the main economic indicators of the farms of 800 ha is presented, in the period 2013–2017, considering the structure, initial production and crop rotation scheme.

Table 3.25: Summary of main indicators of efficiency for module of 800 ha

-lei/expl-

No.	Indicators	2013	2014	2015	2016	2017
1	Value of primary and secondary production	3641760.0	3894528.0	4161984.0	4444128.0	4759200.0
2	Subventions	544000.0	544000.0	544000.0	544000.0	544000.0
3	Gross product (1+2)	4185760.0	4438528.0	4705984.0	4988128.0	5303200.0
4	Total expenses	3113268.8	3261531.2	3439435.2	3617328.0	3795216.0
5	Profit (3–4)	1072491.2	1176996.8	1266548.8	1370800.0	1507984.0
6	Rata of return (%)	32.3	34.1	34.9	36.1	38.0
7	Gross margin for total exploitation	1724718.8	1977486.8	2244942.8	2527086.8	2842158.8
8	Development fund (50% of profit)	536245.6	588498.4	633274.4	685400.0	753992.0
9	Funds for resumption of production (15% of profit)	160873.7	176549.5	189982.3	205620.0	226197.6
10	Funds for capitalization (25% of profit)	268122.8	294249.2	316637.2	342700.0	376996.0

Source: own calculations based on data of Appendix 12

The farm highlights high profitability. The profit rate is high, and it increases from 31.4% in 2013 to 38.0% in 2017. Development Fund established in respect of that exploitation creates the possibility of investment in agricultural equipment (purchase of tractors, combines and other equipment required over the whole range).

Funding for the resumption of production achieved by allocating 15% of the profit and for capitalization (25% profit) provides resetting the process of production, highlighting the ability of the farm to operate viable economic parameters. Setting up a farm of 800 ha may represent an optimal solution in the recovery of the national agricultural potential, considering that the simulations performed, the results confirm not only the economic viability, but the possibility of obtaining economic performance validated by their practical application.

Conclusions

After 1991, Romanian agriculture has been facing extensive processes of fragmentation of agricultural land as a result of the restitution of property to former owners. Structures related to agricultural production and agricultural land were converted into small peasant land ownership as a result of a massive de-structuring of the former production structures. The direct consequences of the agricultural reform have resulted in a rapid increase of the number of landowners and fragmentation of land ownership.

Pronounced fragmentation of agricultural property (land) prevents the achievement of modern agriculture with high yields and acquisition of technical and modern technology.

As a direct, visible and immediate effect of reform measures, there is a decrease in average agricultural area that lies on the farm, which has become one of the lowest in the European Community. Thus, according to European statistics, the average size of a farm in Romania is less than 3.6 times the EU27 average, 1.7 times that of farms in Bulgaria, 1.9 times than those of Hungary, 13 times the holdings in Germany and 14.8 times higher than those in France.

The existence of a large number of small and very small farms contributes to increase structural imbalances in agriculture, with implicit effects on the degree of potential recovery, and it seriously affects the competitiveness of the sector. Small farms are vulnerable to the instability in agricultural markets, having low endowment technology, which contribute to a reduce volume of production or the inability to diversify, while its access to credits is hampered by the lack of means to guarantee. Their sustainable development is hindered by the lack of appropriate policies to support farming.

Decreasing the degree of fragmentation of land ownership cannot be achieved only through the "miracle" of developing a functioning land market, enabling the aggregation of the structures of agricultural production, and hence increasing their economic performance, able to support investment in the purchase of technical and agricultural technology. If we expect the market to solve the disappearance of landownership for 5–8 million people, forcing them to sell their plot of land to save 1–2 years of subsistence, unpredictable consequences may happen.

The market only by itself without coherent and converging interests of the rural population in Romania can still create hideous socio-economic models rejected by economic experiences of the third world.

Farms are the main tools in achieving national agricultural potential capitalization and the improvement of their size and structure is prerequisite to increase the national agricultural potential.

Farms represent the main form of exploitation of national agricultural potential, mobilizing significant resources in the rural communities, in order to exploit the agricultural potential available under a functioning and competitive market economy.

Improving the community agricultural structures was a complex, difficult process that was supported with huge costs financed from public budgets and the budget of Common Agricultural Policy. National financial mechanism, coupled with the result from supporting agricultural policies at EU level, has worked for almost 50 years and favored aggregation of land ownership in the Member States. Financial allocations and support for production represent the factors which have made increase the size of holdings. A large farm offers a higher recovery of involved production factors. Increasing agricultural production could not be achieved only through the operation of large market- oriented farms.

Compliance with the requirements of productivity growth in agriculture and food security has direct effect, but not immediately, gradually decreases in the number of small farms and increases the medium and large ones. In the last 10–15 years, European agriculture has undergone fundamental changes both in terms of financial support and production structures as the result of enlargement of the European Union to 27 Member States and the steps to reform the CAP.

From another perspective, the improvement of agricultural production structures not only meant a continuous scale aggregation and consolidation of large farms, and their diversification, though, the 1992 CAP reform, at the EU level, there is an increase in the number of small farms while the reduction in the number and size of agricultural exploitation is a temperate one. At EU level, large farms are the most important category of farms which realize the bulk of agricultural production. Orientation of agricultural policies by such farms has been done not only because of the necessity to cover the food deficit of the Community, but also for economic reasons: global competition.

Romania's agriculture is highly heterogeneous and polarized in terms of operational structures. Aggregation of land ownership has not been done around large economically viable farms.

The evolution of agricultural production structures in Romania largely reflects the tendency of the owners to preserve the small landed property, regained after 1989. Achieving highly competitive farming structures that correspond to the European agricultural model involves strengthening agricultural areas around the large holdings, able to withstand the market's uncertainties.

Farm size is closely linked to the evolution of agricultural policies, which imposed certain phenomena with impact to the concentration and specialization of agricultural production. Their evolution reflects, in the case of Romania, a slight consolidation of land ownership and building large farms with high productivity. Aggregation of land ownership around large holdings reflects an effort to increase economic efficiency in a sector with traditionally low productivity record.

Despite the measures taken, Romanian agriculture maintains its dual character of farms. It further noted a high fence polarization of farms. Large farms and the individual operating an agricultural area about equal, except that unlike European ones, Romanian individual farms are, mostly, subsistence or semi-subsistence farms, with poor economic results.

Valorization the agricultural potential in large farms is a solution able to meet the market requirements, through the volume and structure of production. Considering, however, the specificity of this sector, addressing the entire issue relating to the agricultural potential in the context of integration and globalization of markets must be done in a much broader context.

Romania's integration in the economic community framework and applying CAP mechanisms significantly influence the evolution of the number and structure of agricultural holdings of national agricultural policy measures taken at the level of each individual Member State and EU-27. Financial measures to support farm will contribute significantly to reducing the reluctance to ways of association in Romania and establishing agricultural producer groups, and as an intrinsic effect there can be noticed a consolidation of farming forms.

Small peasant holding can become competitive given its focus on multi-functional, multiple activities and non-agricultural income diversification. Individual farm is an exponent of the transformations and realities of national agricultural model, having an important role in the rural economy as a whole, through the social dimension that implies. The individual farm simultaneously unites the individual household agricultural production capacities, the family and their means of production, thus helping to maintain and achieve a functional balance in rural areas.

Based on the realities of the European agricultural model to which the Romanian agriculture tends to integrate where the small and medium farms hold the dominant role and receive high financial support, Romania can capitalize on the diversity of the small farms and increase their multi-functional role in ensuring security for a significant proportion of the rural population.

Subsistence farming practiced in small peasant holdings contributes significantly to improving the material situation of families in rural areas, acting, as some specialized studies have remarked (Kostov and Lingard, 2002), 30 as protection against absolute poverty, by ensuring a minimum level of food and income.

Small peasant farm is an essential component in the Romanian agricultural economy that has significant share in overall agricultural production structures. In Romania, according to statistics (National Institute of Statistics, 2006), about 3 million farms do not meet the minimum requirements to be included in farms category (they do not hold an area of 1 ha of agricultural land and plots> 0.3 ha) and only 29% the total number of agricultural households (1.24 million farms in 2007 were registered in the Land Register of farms).31

For rural communities, subsistence farming can be the insurance against economic risks - even if a precarious one, as consider Abele

30 Philip Kostov, John Lingard (2002), *Subsistence farming in transitional economies: lessons from Bulgaria*, Journal of Rural Studies, 18(1): 83–94

31 ***INS (2006), *Structura agriculturii româneşti, în pragul aderării României la Uniunea Europeană*, Comunicat de presă, 62/7 decembrie, Bucureşti, www.insse.ro/cms/files/statistici/comunicate/alte/comunicat%20ASA.pdf

and Forsberg (2003)[32] consider, under deepening economic crisis and economic imbalances in the national economy. Agricultural production obtained in peasant farms often provides the only survival means of small rural households and achieves a substantial part of their income, but it was ineffective in harnessing the potential of national agricultural exploitation.

Development of multifunctional agriculture and farm income diversification, especially of the non-agriculture one, can be a viable solution in improving the competitiveness of farms in Romania especially that they have and exploit an important tourist and ecological potential. The solution might be to adapt to the realities of Romanian agriculture the model of small French farms by increasing and diversifying agro-tourist activities provided within individual farms; production and supply organic products with high added value and label of origin; combining farming in holding with activities out of it.

Implementation and adaptation the measures outlined to the Romanian realities of subsistence agricultural model will generate positive effects in terms of awareness the need to develop a multifunctional agriculture in Romania, in which the peasant small holdings to capitalize not only agricultural land but also experiences and cultural values of rural areas and their agro-tourist potential.

Recent developments in the economy and the community have demonstrated the need to develop strategies that require the identification and implementation of adaptation measures to the potential of Romanian agriculture in economic and social realities in terms of economic instability increasingly apparent.

The recovery of national potential of agriculture should be a major concern for all stakeholders involved, since due to its size the agricultural sector is an important resource in the highly competitive construction of market economy in Romania. The orientation of this sector to the values of the market economy, beyond the social role and support that it still

32 Steffen Abele, Klaus Frohberg (2003), *Subsistence Agriculture in Central and Eastern Europe: How to Break the Vicious Circle?*, Studies on the Agricultural and Food Sector in Central and Eastern Europe, Vol. 22, Leibniz Institute of Agricultural Development in Central and Eastern Europe (IAMO), Halle, p. IV

has for rural population, can contribute significantly to strengthening the financial resources to improve the contribution to the GDP.

Starting from the two existing concrete realities of the Romanian agriculture, respectively pronounced fragmentation of land ownership and the existence of an oversize rural agricultural population, the first steps should focus on finding solutions to stimulate and strengthen farm land consolidation through cooperation and training producer groups, in the framework of agricultural diversification which must mobilize the surplus of agricultural population.

The existence of agricultural overpopulation is one of the major problems faced by Romanian agriculture and whose outcome depends largely on developments in rural areas and the development of multifunctional agriculture in rural communities.

From this perspective, a holistic approach to rural is required, particularly aiming to improve agricultural production structures and the orientation of their production to the market, thus increasing the rate of absorption of labor in agricultural production structures to build a multifunctional rural environment. Improving farm structure may turn primarily to the correct application of the law of the land and legal circulation of land, to facilitate procedural consolidation of farms and land consolidation. To all these there must be added amenities to stimulate the appearance of producer groups, to supervise the application of funding schemes for young farmers and build new farms in rural areas.

To create the conditions for the functioning of a modern and competitive agriculture also requires the existence of specific services for agriculture, providing means needed for land resources recovery at affordable prices for farmers, given the high costs of purchasing agricultural equipment, which makes impossible to apply new technologies. In Romania, there should be established operators to provide integrated farming, starting perhaps from experience of former SMAs, before 1989, in agricultural services and ending with the network collection - transport - processing.

The lack of professional services for collecting and transporting agricultural products to processors and interposing the commercial chain of unprofessional intermediaries significantly reduce farmers' income and thus decrease the ability to achieve additional agricultural production and ultimately increase the price of agricultural products.

Increasing the efficiency of the agricultural sector can be achieved by strengthening a performing network of agricultural loan that should provide cheap financial resources for the development of agricultural production structures both by stimulating the purchase of agricultural machinery and especially the acquisition of new land and participation in the European co-financing programs for farmers. Promotion loans with subsidized interest rates and payment appropriations committed by the overall production capacity expansion can stimulate agricultural production and start new activities that will help improve employment in agriculture and modernization of rural communities.

Promoting multifunctional vision, where agriculture is no longer the only income generating activity for rural communities, but rather would be combined with complementary activities such as agro tourism and rural small industry would be a fundamental dimension in revitalizing the rural Romanian environment.

Analysis performed in the Romanian agriculture reveals that the achievement of modern and competitive agriculture, which recovers the potential of agriculture to a higher level, cannot be achieved without understanding the intimate operation mechanisms and correlations between agriculture and national economy.

To identify the best ways to increase the recovery of national agricultural potential will be, at least for some time, one of the major concerns of stakeholders in the development of Romanian agro-food system, given on the one hand, the massive fragmentation of land ownership, agricultural overpopulation and pronounced ruralization.

Bibliography

Adrian Turek Rahoveanu, Magdalena Turek Rahovean, Dinu Toma (2011), *Optimization of production structures in order to increase competitiveness of agricultural holdings*, Economics of agriculture, (LVIII): 187–194.

Agata Pieniadz, Jon Hanf, Dan Marius Voicilas, Stefan Wegener, Linde Götz (2009), *Small farmers in the Romanian dairy market: Do they have a future?*, The 111 EAAE-IAAE Seminar 'Small Farms: decline or persistence' University of Kent, Canterbury, 26–27 June, UK.

Ajay Singh, Sudhindra Nath Panda (2012), *Development and application of an optimization model for the maximization of net agricultural return*, Agricultural Water Management, (115): 267– 275.

Alexandru Tofan (2005/2006), *Dimensiunea economică a exploatațiilor agricole*, Analele Ştiinţifice Ale Universităţii „Alexandru Ioan Cuza" Din Iaşi, Tomul LII/LII Ştiinţe Economice 2005/2006.

Anca Dachin (2011), *Contribuții ale agriculturii la fluctuațiile economice în România*, Economie teoretică şi aplicată, Vol. XVIII, 1(554): 154–165.

Anderson T.W., Cheng Hsiao (1982), *Formulation and estimation of dynamic model using panel data*, Journal of Econometrics (18): 47–82.

Andrei Jean, Dusmanescu Dorel (2012), *Some possible effects of Common Agricultural Policy (CAP) on Romanian Agriculture during the 2014–2020 financial perspective. How much does the paradigm really change?*, Proceedings of the 3rd International Symposium Agrarian Economy and Rural Development (AERD'12) "Agrarian economy and rural development - realities and perspectives for Romania", October 11–13, Bucharest, Romania, p. 8–15.

Bogdan Murgescu (2010), *România şi Europa. Acumularea decalajelor economice (1500–2010)*, Editura Polirom, Iaşi.

Bran Mariana, Dobre Iuliana, Stefan Marcela, Voicu Radu (2011), *Opportunities for the vegetal biodiversity - support for the sustainable development in the South Muntenia region,* Journal Quality – Access to Success, 12(121): 32–36.

Calcedonia Enache (2010), *The utilization of the statistical techniques in projecting gross value added in the agriculture, hunting and forestry; fishery and pisciculture sector*, Agricultural Economics and Rural Development, 7(2): 285–291.

Camelia Burja, Vasile Burja (2008), *Performanţa economică a exploataţiilor agricole în sistemul dezvoltării durabile*, Editura Cărţii de Ştiinţă, Cluj-Napoca.

Carmen Bucur (2012), *The Romanian agri-food aggregate demand – evolutions and trends in the context of the economic and financial crisis*, Agricultural Economics and Rural Development, 9(1): 71–83.

Cecilia Alexandri (2010), *Farm consolidation – first signal after accession*, Agricultural Economics and Rural Development, VII (2): 239–248.

Constantin Ciutacu, Luminita Chivu, Mark Carley (2008), *E'tude comparative dans le secteur de l'agriculture de L'union Européenne*, Romanian Journal of Economics, 27/2(36): 40–56.

Crina Turtoi, Camelia Toma, Camelia Gavrilescu (2007), *Changes in the Romanian farm structures during transition – evolution and main determinants*, in Proceedings of Agricultural economics and transition: What was expected, what we observed, the lessons learned. (Volume I), Leibniz Institute of Agricultural Development in Central and Eastern-Europe (IAMO), Halle (Saale).

Crina Turtoi, Camelia Toma, Camelia Gavrilescu (2008), *Changes in the Romanian farm structures during transition – evolution and main determinants*, in Proceedings of Agricultural economics and transition: What was expected, what we observed, the lessons learned. (Volume I), Leibniz Institute of Agricultural Development in Central and Eastern-Europe (IAMO), Halle (Saale).

Cristian Socol (2004), *Opţiunile de politică agricolă şi dezvoltare durabilă a agriculturii*, Editura Economică, Bucureşti.

Eniko Veress (1999), *De/reţărănizare în România după 1899*, Sociologie Românească, (3): 79–90.

Filon Toderoiu (2012), *The Romanian agri-food economy – performance reductive effects after five years of EU membership*, Agricultural Economics and Rural Development, 9(1): 25–45.

Filon Toderoiu (2011), *Macroeconomic variables influencing the European convergence of the Romanian agri-food sector*, Agricultural Economics and Rural Development, 8(1): 13–43

Filon Toderoiu (2004), *Romanian agrifood sector-macroeconomic correlations of domestic competitiveness*, Journal for Economic Forecasting, 1(4): 87–105.

Filon Toderoiu, Camelia Stefanescu (2002), *Agri-food economy in Romania: loss of internal and external competitiveness*, Journal for Economic Forecasting, (1): 62–75

Hironori Yagi, Guy Garrod (2006), *An empirical application of the linear programming model to agricultural land-use planning: a valuation of externalities on farmland plots from unused land*, International Journal of Biodiversity Science & Management, (2:3): 227–231.

Holtz-Eakin Douglas, Newey Whitney, Rosen Harvey (1988), *Estimating vector autoregressions with panel data*. Econometrica 56, 1371–1395.

Ioan Done, Luminita Chivu, Jean Andrei, Mirela Matei (2012), *Using labor force and green investments in valuing the Romanian agriculture potential*, Journal of Food, Agriculture & Environment Vol. 10 (3&4): 737–741.

Ioan Done (2011), *The Economic Welfare between the Fundamental Economic Correlations and the Harmonization of the Company's Interests (I)*, Bulletin of Petroleum-Gas University of Ploiesti, Economic Sciences Series, Vol. LXIII. (2): 105–114.

Jean Andrei, Dorel Dusmanescu (2012), *Some Romanian experience in achieving the best Common Agricultural Policy results. How much has the agricultural sector has beneficiate?*, in Proceedings of The International Scientific Meeting „Sustainable agriculture and rural development in terms of the Republic of Serbia strategic goals realization within the Danube region - preservation of rural values", December, 6–8th, Tara, Serbia, p. 801–819

Jean Andrei, Dorel Dusmanescu, Marius Voicilas (2011), *Common Market Organization a new challenge in reforming the Common Agricultural Policy*, Economics of Agriculture, Vol. LVIII CSI-1 (1–412), 58(3 special): 271–278.

Jean Andrei, Mircea Untaru (2012), *Through Common Agricultural Policy Reforms: A Short Analysis*, International Journal of Sustainable Economies Management, 1(3): 32–40.

Zhaorong Jiao, Andrew Higgins, Diane Prestwidge (2005), *An integrated statistical and optimization approach to increasing sugar production within a mill region*, Computers and Electronics in Agriculture, (48): 170–181.

Johan Swinnen, Liesbet Vranken (2010), *Review of the transitional restrictions maintained by Bulgaria and Romania with regard to the acquisition of agricultural real estate* (Final Report), Centre for European Policy Studies, Brussels.

Johannes Sauer, Borbala Balint (2006), *Small-scale farming in Romania –shadow prices and efficiency*, Studies on the Agricultural and Food Sector in Central and Eastern Europe, Vol. 33, IAMO, Halle (Saale),

John Colinsk (1989), *An Aggregate Model of Technical Change*. The Quarterly Journal of Economics, 104(4): 787–821.

John von Neumann (1954), *A numerical method to determine optimum strategy*. Naval Research Logistics Quarterly, 1(2): 109–115.

Jonel Subic, Jean Andrei, Ion Iarca, Dorel Dusmanescu, Ioan Done (2011), *Optimal Allocation of Investment Resources for Agricultural Holding Modernization Using the Linear Programming*, **Bulletin of University of Agricultural Sciences and Veterinary Medicine Cluj-Napoca. Horticulture**, 68(2): 217–224.

Karlheinz Knickel, Gianluca Brunori, Sigrid Rand, Jet Proost (2009), *Towards a Better Conceptual Framework for Innovation Processes in Agriculture and Rural Development: From Linear Models to Systemic Approaches*, The Journal of Agricultural Education and Extension, (15:2): 131–146.

Lucian Luca, Cristina Cionga, Daniela Giurcă (2012), *Consolidarea exploatațiilor agricole*, Editura Economică, Bucureşti

Lucian Luca (2007), *Romania: Large Semi-subsitence Farm Sector, Result of Wrong Strategic Approach?,* The 104th (joint) EAAE-IAAE Seminar Agricultural Economics and Transition: What was expected, what we observed, the lessons learned, Corvinus University of Budapest, 6–8 September Hungary.

Luminiţa Chivu (2002), *Competitivitatea în agricultură – analize şi comparaţii europene*, Editura Expert, Bucureşti.

Luminiţa Chivu, Constantin Ciutacu, Valeriu Ioan-Franc (2003), *Agricultura între restricţiile comerciale globale şi politicile comunitare*, Centrul de Economie Comparată şi Consens, Academia Română, Editura Expert, Bucureşti.

Mălina Voicu, Bogdan Voicu (2006), *Satul Românesc pe drumul către Europa*, Editura Polirom, Iaşi

Manea Drăghici, Livia David, Valentina Tudor (2010), *Facilities of using linear programing for structure optimization Of the agricultural holdings activities*, Scientific Papers Management, Economic Engineering in Agriculture and Rural Development, 10(2): 71–75.

Nevo A., Oad R., Podmore T. (1994), *An integrated expert system for optimal crop planning*, Agricultural Systems (45): 73–92.

agriculture - the gap between the EU-27, Economics of agriculture, (2): 366–374.

Radu Voicu, Iuliana Dobre (2003), *Organizarea şi strategia dezvoltării unităţilor agricole*, Editura, ASE, Bucureşti.

Raju K.S, Kumar D.N. (1999), *Multicriterion decision-making in irrigation planning*, Agricultural Systems (62): 117–129.

Rizov Marian, Dinu Gavrilescu, H.Gow, Erik Mathijs, Johan Swinnen (1999), *Transition and Enterprise Restructuring: the Development of Individual Farming in Romania*, EU Phare ACE and the World Bank/ECSSD/DECRG Working Paper no. 3 December.

Robert Ayres, Leslie Ayres, Ben Warr (2002), *Energy, Power and Work in the U. S. Economy 1900–1998*. INSEAD'S Center For the Management of Environmental Resources, 2002/52/EPS/CMER.

Romică Trandafir (2004), *Modele şi algoritmi de optimizare*, Editura Agir, Bucureşti.

Ruhul Sarker, Saifullah Talukdar, Anwarul Haque (1997), *Determination of optimum crop mix for crop cultivation in Bangladesh*, Appl. Math. Modelling, (21): 621–632.

Rusovici Alexandru (1991), *Agricultura în economia de piaţă*, Editura Ceres, Bucureşti.

Sabates Wheeler (2002), Farm Strategy, Self-Selection and Productivity: *Can Small Farming Groups Offer Production Benefits to Farmers in Post-Socialist Romania?*, *World Development*, 30(10): 1737–1753.

Sangho Kim, Hyunjoon Lim (2004), *Does the Solow Residual for Korea Reflect Pure Technology Shocks?*. Econometric Society 2004 Far Eastern Meetings

Stefan Tangermann (2011), *Direct payments in the CAP post 2013*, European Parliament's Committee on Agriculture and Rural Development, Brussels.

Terry Marsden (1999), *Rural futures: the consumption of the countryside and its regulation*. Sociologia Ruralis (39): 501–520.

Tudorel Andrei, Bourbonnais Regis (2008), *Econometrie*. Editura Economică, Bucureşti.

Turgutlu Evrim (2010), *Cyclical behavior of price-cost margins in the Turkish banking industry*. Economic Modelling (27): 368–374.

Xieting Zeng, Shaozhong Kang, Fusheng Li, Lu Zhang, Ping Guo (2010), *Fuzzy multi-objective linear programming applying to crop area planning*, Agricultural Water Management (98): 134–142.

***Comisia Prezidenţială pentru Politici Publice de Dezvoltare a Agriculturii (2013), *Cadrul naţional strategic pentru dezvoltarea durabilă a sectorului agroalimentar şi a spaţiului rural în perioada 2014–2020–2030*, Bucureşti, http://www.acad.ro/forumuri/pag_forum_2013_0701Stategie.htm.

***European Commission (2011), *Common Agricultural Policy towards 2020. Assessment of Alternative Policy Options,* SEC (2011) 1153 final/2, Brussels, http://ec.europa.eu/agriculture/policy-perspectives/impact-assessment/cap-towards-2020/index_en.htm#ia.

***European Commission, Directorate General for Agriculture and Rural Development (2012), *Agriculture in the European Union. Statistical and economic information 2012,* European Union, Luxembourg, 2012.

***European Commission, Directorate General for Agriculture and Rural Development (2012), *Rural Development in the European Union - Statistical and economic information - 2012*, European Union, Luxembourg, 2012

*** European Commission, Directorate General for Agriculture and Rural Development (2013), *Agricultural Policy Perspectives. Member States factsheets - May 2013. Romania*, European Union, 2013.

***EUROSTAT (2012), *Agriculture and fishery statistics. Main results 2010–2011*, Luxembourg.

***EUROSTAT (2012), *Eurostat yearbook 2012. Europe in figure*, Luxembourg.http://epp.eurostat.ec.europa.eu/statistics_explained/index.php/Europe_in_figures_-_Eurostat_yearbook

***INS (2011), *Anuarul Statistic al României. Serie de date 1990–2010*, Institutul Naţional de Statistică, Bucureşti

***INS (2011), *Economia mondială în cifre - 2011,* Institutul Naţional de Statistică, Bucureşti,

***INS (2012), *Recensământul General Agricol 2010*, Institutul Naţional de Statistică, Bucureşti

***INS (2013), *Anuarul Statistic al României 2012*, Institutul Naţional de Statistică, Bucureşti

Appendixes

Appendix 1

Crop rotation for a 60 ha farm in a lowland agricultural area

Year/Sole	I	II	III	IV	V	VI	VII
2012	Wheat 13 ha	Barley 7 ha	Corn 17 ha	Sunflower 10 ha	Rape 6 ha	Sugar Beet 3 ha	Beans 4 ha
2013	Corn 3 ha Sunflower 10 ha	Corn 7 ha	Corn 4 ha Sugar Beet 3 ha Beans 4 ha Rape 6 ha	Wheat 3 ha Barley 7 ha	Wheat 6 ha	Corn 3 ha	Wheat 4 ha
2014	Wheat 13 ha	Barley 7 ha	Corn 17 ha	Sunflower 10 ha	Rape 6 ha	Sugar Beet 3 ha	Beans 4 ha
2015	Corn 3 ha Sunflower 10 ha	Corn 7 ha	Corn 4 ha Sugar Beet 3 ha Beans 4 ha Rape 6 ha	Wheat 3 ha Barley 7 ha	Wheat 6 ha	Corn 3 ha	Wheat 4 ha
2016	Wheat 13 ha	Barley 7 ha	Corn 17 ha	Sunflower 10 ha	Rape 6 ha	Sugar Beet 3 ha	Beans 4 ha

Source: author based on constructive assumptions

Basically details in building a 60 ha farm in a lowland agricultural area

Surfaces and the average yield of crops in crop rotation of 60 ha

Crop	Share culture in rotation (%)-	The allocated area (ha)	Average agricultural production
Wheat	21,7%	13 ha	3000 kg/ha N
Barley	11,7%	7 ha	3200 kg/ha N
Corn	28,3%	17 ha	3200 kg/ha N
Sunflower	16,7%	10 ha	1300 kg/ha N
Rape	10%	6 ha	1600 kg/ha N
Sugar Beet	5%	3 ha	20000 kg/ha N
Beans	6,6%	4 ha	800 kg/ha N

Note: N=non-irrigated
Source: author's proposal

Share culture in rotation(%)

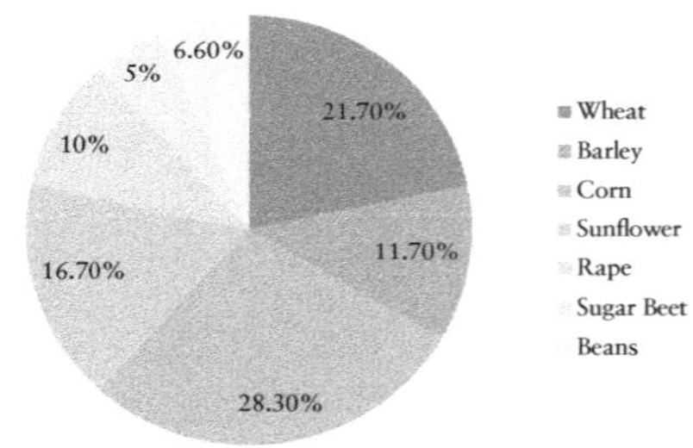

Appendix 2

Crop rotation for a 100 ha farm in a lowland agricultural area

Year/Sole	I	II	III	IV	V	VI	VII
2012	Wheat 13 ha	Barley 7 ha	Corn 17 ha	Sunflower 10 ha	Rape 6 ha	Sugar Beet 3 ha	Beans 4 ha
2013	Corn 3 ha Sunflower 10 ha	Corn 7 ha	Corn 4 ha Sugar Beet 3 ha Beans 4 ha Rape 6 ha	Wheat 3 ha Barley 7 ha	Wheat 6 ha	Corn 3 ha	Wheat 4 ha
2014	Wheat 13 ha	Barley 7 ha	Corn 17 ha	Sunflower 10 ha	Rape 6 ha	Sugar Beet 3 ha	Beans 4 ha
2015	Corn 3 ha Sunflower 10 ha	Corn 7 ha	Corn 4 ha Sugar Beet 3 ha Beans 4 ha Rape 6 ha	Wheat 3 ha Barley 7 ha	Wheat 6 ha	Corn 3 ha	Wheat 4 ha
2016	Wheat 13 ha	Barley 7 ha	Corn 17 ha	Sunflower 10 ha	Rape 6 ha	Sugar Beet 3 ha	Beans 4 ha

Source: authors based on constructive assumptions

Basically details in building a 100 ha farm in a lowland agricultural area

Surfaces and the average yield of crops in crop rotation of 100 ha

	Share culture in rotation -(%)-	The allocated area	Average agricultural production
Wheat	22%	22 ha	3200 kg/ha N
Barley	11%	11 ha	3400 kg/ha N
Corn	22%	22 ha	3400 kg/ha N
Sunflower	16%	16 ha	1400 kg/ha N
rape	12%	12 ha	1800 kg/ha N
Sugar Beet	5%	5 ha	23000 kg/ha N
Beans	12%	12 ha	1000 kg/ha N

Note: N=non-irrigated
Source: author's proposal

Share culture in rotation(%)

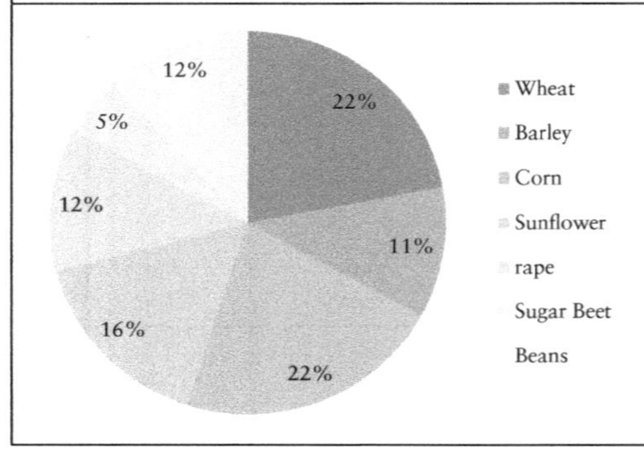

Appendix 3

Crop rotation for a 250 ha farm in a lowland agricultural area

Year/Sole	I	II	III	IV	V	VI
2012	Wheat 55 ha	Barley 25 ha	Corn 50 ha	Sunflower 40 ha	Rape 30 ha	Soy 50 ha
2013	Rape 15 ha Sunflower 40 ha	Rape 15 ha Soy 10 ha	Soy 40 ha Corn 10 ha	Wheat 25 ha Barley 15 ha	Wheat 30 ha	Corn 40 ha Barley 10 ha
2014	Wheat 55 ha	Barley 25 ha	Corn 50 ha	Sunflower 40 ha	Rape 30 ha	Soy 50 ha
2015	Rape 15 ha Sunflower 40 ha	Rape 15 ha Soy 10 ha	Soy 40 ha Corn 10 ha	Wheat 25 ha Barley 15 ha	Wheat 30 ha	Corn 40 ha Barley 10 ha
2016	Wheat 55 ha	Barley 25 ha	Corn 50 ha	Sunflower 40 ha	Rape 30 ha	Soy 50 ha

Source: authors based on constructive assumptions

Basically details in building a 250 ha farm in a lowland agricultural area

Surfaces and the average yield of crops in crop rotation of 250 ha

Crop	Share culture in rotation -(%)-	The allocated area	Average agricultural production
Wheat	22%	55 ha	3500 kg/ha N
Barley	10%	25 ha	3600 kg/ha N
Corn	20%	50 ha	3600 kg/ha N
Sunflower	16%	40 ha	1600 kg/ha N
Rape	12%	30 ha	2000 kg/ha N
Soy	20%	50 ha	2000 kg/ha N

Note: N=non-irrigated
Source: author's proposal

Share culture in rotation(%)

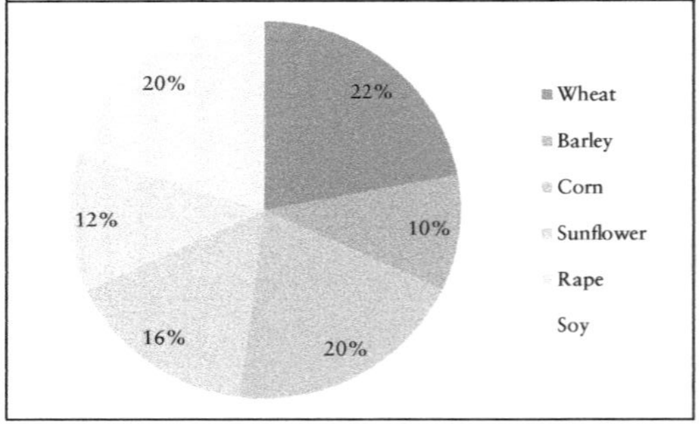

Appendix 4

Crop rotation for a 500 ha farm in a lowland agricultural area

Year/Sole	I	II	III	IV	V	VI
2012	Wheat 100 ha	Barley 50 ha	Corn 100 ha	Sunflower 80 ha	Rape 70 ha	Soy 100 ha
2013	Sunflower 80 ha Rape 20 ha	Rape 50 ha	Soy 100 ha	Wheat 30 ha Barley 50 ha	Wheat 70 ha	Corn 100 ha
2014	Wheat 100 ha	Barley 50 ha	Corn 100 ha	Sunflower 80 ha	Rape 70 ha	Soy 100 ha
2015	Sunflower 80 ha Rape 20 ha	Rape 50 ha	Soy 100 ha	Wheat 30 ha Barley 50 ha	Wheat 70 ha	Corn 100 ha
2016	Wheat 100 ha	Barley 50 ha	Corn 100 ha	Sunflower 80 ha	Rape 70 ha	Soy 100 ha

Source: authors based on constructive assumptions

Basically details in building a 500 ha farm in a lowland agricultural area

Surfaces and the average yield of crops in crop rotation of 500 ha

Crop	Share culture in rotation -(%)-	The allocated area	Average agricultural production
Wheat	20%	100	3800 kg/ha N
Barley	10%	50 ha	4000 kg/ha N
Corn	20%	100 ha	4000 kg/ha I
Sunflower	16%	80 ha	1800 kg/ha N
Rape	14%	70 ha	2200 kg/ha N
Soy:	20%	100 ha	2200 kg/ha I

Note: I=irrigated N=non-irrigated
Source: author's proposal

Share culture in rotation(%)

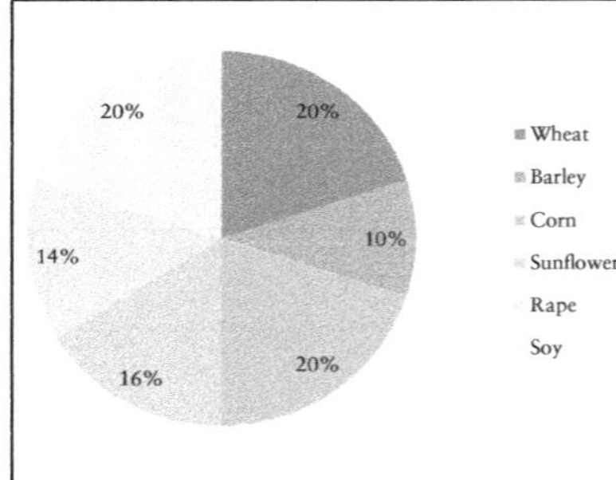

Appendix 5

Crop rotation for a 800 ha farm in a lowland agricultural area

Year/Sole	I	II	III	IV	V	VI
2012	Wheat 160 ha	Barley 80 ha	Corn 160 ha	Sunflower 128 ha	Rape 112 ha	Soy 160 ha
2013	Sunflower 128 ha Rape 32 ha	Rape 80 ha	Soy 160 ha	Barley 80 ha Wheat 48 ha	Wheat 112 ha	Corn 160 ha
2014	Wheat 160 ha	Barley 80 ha	Corn 160 ha	Sunflower 128 ha	Rape 112 ha	Soy 160 ha
2015	Sunflower 128 ha Rape 32 ha	Rape 80 ha	Soy 160 ha	Barley 80 ha Wheat 48 ha	Wheat 112 ha	Corn 160 ha
2016	Wheat 160 ha	Barley 80 ha	Corn 160 ha	Sunflower 128 ha	Rape 112 ha	Soy 160 ha

Source: authcrs based on constructive assumptions

Basically details in building a 800 ha farm in a lowland agricultural area

Surfaces and the average yield of crops in crop rotation of 800 ha

Crop	Share culture in rotation -(%)-	The allocated area	Average agricultural production
Wheat	20%	160 h	3800 kg/ha N
Barley	10%	80 ha	4000 kg/ha N
Corn	20%	160 ha	4000 kg/ha I
Sunflower	16%	128 ha	1800 kg/ha N
rape	14%	112 ha	2200 kg/ha N
Soy	20%	160 ha	2400 kg/ha I

Note: I=irrigated N=non-irrigated
Source: author's proposal

Share culture in rotation(%)

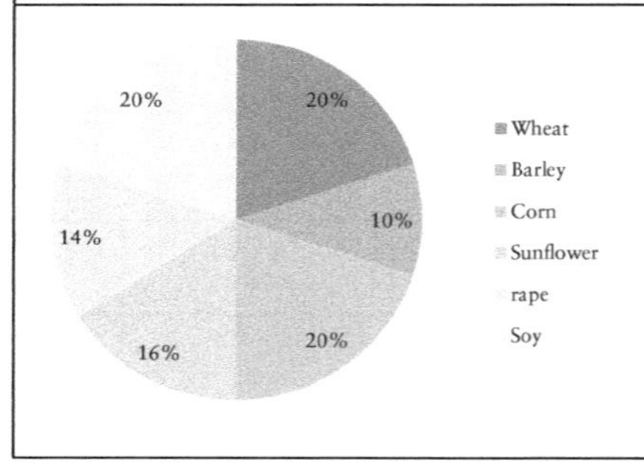

Appendix 6

Farms by economic types and size in Romania

The classification of farms by economic size classes

Class	Limits IN UDE	Class name
I	<2	Very small
II	>=2 and<4	
III	>=4 and<6	Small
IV	>=6 and <8	
V	>=8 and<12	Small medium
VI	>=12 and<16	
VII	>=16 and<40	Medium large
VIII	>=40 and<100	Big
IX	>=100 and<250	Large
X	>=250	Extra

Source: based on 85/377/CEE and 1999/725/CE) 22

Appendix 7

Agricultural products prices in agri-food market, 2006–2017

-lei/kg-

Products year	Wheat	Barley	Corn	Sunflower	Soy	Sugar beet	Rape	Beans
2006	0,36	0,35	0,31	0,72	0,63	0,07	0,60	3,16
2007	0,44	0,36	0,30	0,70	0,57	0,09	0,74	3,43
2008	0,61	0,59	0,45	0,84	0,78	0,10	0,79	3,21
2009	0,66	0,67	0,52	0,92	0,97	0,11	1,00	3,37
2010	0,77	0,74	0,58	0,96	1,06	0,18	1,07	3,96
2011	0,79	0,71	0,67	1,19	1,23	0,22	1,25	3,38
2012	1,00	0,85	0,90	1,50	2,10	0,30	1,50	3,50
2013*	1,07	0,90	0,96	1,60	2,24	0,32	1,60	3,74
2014*	1,14	0,97	1,03	1,71	2,40	0,34	1,71	4,00
2015*	1,22	1,04	1,10	1,83	2,57	0,36	1,83	4,28
2016*	1,31	1,11	1,17	1,96	2,75	0,39	1,96	4,58
2017*	1,40	1,19	1,26	2,10	2,94	0,42	2,10	4,90

Observation: The prices form 2006–2011 represents the average prices on the peasant market based on farm survey in South Region.
The prices from 2013*–2017* are predicted
Source: author's own computation

Appendix 8

Annual Budget For The Holding Of 60 Ha, 2013–2017

Indicators	U.M	2012		2013		2014		2015		2016		2017	
		lei/expl.	€/expl.	lei/expl.	€/expl.	lei/expl.	€/expl.	lei/expl.	€/expl.	lei/expl.	€/expl.	lei/expl.	€/expl.
A. PRODUCTION VALUE	lei	173320,0	40307,0	184662,0	42944,7	197286,0	45880,5	210590,0	48974,4	225174,0	52366,0	241360,0	56130,2
A1. From the main production	lei	170100,0	39558,1	181442,0	42195,8	194066,0	45131,6	207370,0	48225,6	221954,0	51617,2	238140,0	55381,4
B (+) SUBSIDIES	lei	40800,0	9488,4	40800,0	9488,4	40800,0	9488,4	40800,0	9488,4	40800,0	9488,4	40800,0	9488,4
C (=) GROSS PRODUCT	lei	214120,0	49795,3	225462,0	52433,0	238086,0	55368,8	251390,0	58462,8	265974,0	61854,4	282160,0	65618,6
D (-) TOTAL EXPENSES	lei	182537,6	42450,6	191661,0	44572,3	200786,9	46694,6	211741,5	49242,2	222700,0	51790,7	233643,9	54335,8
D1. From the main production	lei	182534,4	42449,9	191657,8	44571,6	200783,7	46693,9	211738,3	49241,5	222696,8	51789,9	233640,7	54335,0
I. VARIABLE COSTS	lei	153192,5	35626,2	160852,1	37407,5	168511,7	39188,8	177703,3	41326,3	186894,8	43463,9	196086,4	45601,5
1. Raw materials and supplies expenditure	lei	68726,1	15982,8	72162,4	16781,9	75598,7	17581,1	79722,3	18540,1	83845,8	19499,0	87969,4	20458,0
- Seedlings	lei	24923,4	5796,1	26169,6	6085,9	27415,7	6375,8	28911,1	6723,5	30406,5	7071,3	31902,0	7419,1
- Fertilizers	lei	22742,4	5288,9	23879,5	5553,4	25016,6	5817,8	26381,2	6135,2	27745,7	6452,5	29110,3	6769,8
- Pesticides	lei	19459,0	4525,3	20432,0	4751,6	21404,9	4977,9	22572,4	5249,4	23740,0	5520,9	24907,5	5792,4
- Other materials	lei	1601,3	372,4	1681,3	391,0	1761,4	409,6	1857,5	432,0	1953,6	454,3	2049,6	476,7
2. Expenses with mechanical works	lei	77813,1	18096,1	81703,8	19000,9	85594,4	19905,7	90263,2	20991,4	94932,0	22077,2	99600,8	23163,0
3. Expenditure on irrigation	lei	0,0	0,0	0,0	0,0	0,0	0,0	0,0	0,0	0,0	0,0	0,0	0,0
4. Supply Expenses	lei	3436,3	799,1	3608,1	839,1	3779,9	879,1	3986,1	927,0	4192,3	975,0	4398,5	1022,9

Indicators	U.M	2012		2013		2014		2015		2016		2017	
		lei/expl.	€/expl.	lei/expl.	€/expl.	lei/expl.	€/expl.	lei/expl.	€/expl.	lei/expl.	€/expl.	lei/expl.	€/expl.
5. Temporary labor expenses	lei	0,0	0,0	0,0	0,0	0,0	0,0	0,0	0,0	0,0	0,0	0,0	0,0
6. Insurance	lei	3217,0	748,1	3377,9	785,5	3538,7	823,0	3731,7	867,8	3924,7	912,7	4117,8	957,6
II. FIXED COSTS	lei	29345,1	6824,4	30808,9	7164,9	32275,2	7505,9	34038,2	7915,9	35805,2	8326,8	37557,5	8734,3
- Permanent labor expenses	lei	10874,8	2529,0	11418,5	2655,5	11962,3	2781,9	12614,7	2933,7	13267,2	3085,4	13919,7	3237,1
- Overheads	lei	4825,5	1122,2	5066,8	1178,3	5308,1	1234,4	5597,6	1301,8	5887,1	1369,1	6176,7	1436,4
- Interest	lei	9444,8	2196,5	9917,1	2306,3	10389,3	2416,1	10956,0	2547,9	11522,7	2679,7	12089,4	2811,5
- Depreciation	lei	4200,0	976,7	4406,5	1024,8	4615,5	1073,4	4869,9	1132,5	5128,1	1192,6	5371,7	1249,2
E. (=) TAXABLE INCOME	lei	-12434,4	-2891,7	-10215,8	-2375,8	-6717,7	-1562,3	-4368,3	-1015,9	-742,8	-172,7	4499,3	1046,4
(-) Taxes	lei	0,0	0,0	0,0	0,0	0,0	0,0	0,0	0,0	0,0	0,0	0,0	0,0
F. (=)NET REVENUE+SUBSIDIES	lei	28365,6	6596,7	30584,2	7112,6	34082,3	7926,1	36431,7	8472,5	40057,2	9315,6	45299,3	10534,7
G. RATE OFTAXABLE INCOME	%	-6,8	-6,8	-5,3	-5,3	-3,3	-3,3	-2,1	-2,1	-0,3	-0,3	1,9	1,9
H. RATE OF NET REVENUE +SUBSIDIES	%	15,5	15,5	16,0	16,0	17,0	17,0	17,2	17,2	18,0	18,0	19,4	19,4

Source: author's own computation

Appendix 8a

BUDGET OF 60 ha FARM IN LOWLAND AREA
Basic variant – year 2012

Indicators	U.M	CROP							TOTAL	
		Wheat	Barley	Corn	Sunflower	Rape	Sugar beet	Beans	lei/expl.	€/expl.
		ha								
		13,00	7,00	17,00	10,00	6,00	3,00	4,00	60,00	60,00
		21,7%	11,7%	28,3%	16,7%	10,0%	5,0%	6,6%	100%	100%
A. PRODUCTION VALUE	lei	39520,0	19600,0	50320,0	19850,0	14670,0	18000,0	11360,0	173320,0	40307,0
A1. From the main production	lei	39000,0	19040,0	48960,0	19500,0	14400,0	18000,0	11200,0	170100,0	39558,1
B (+) SUBSIDIES	lei	8840,0	4760,0	11560,0	6800,0	4080,0	2040,0	2720,0	40800,0	9488,4
C (=) GROSS PRODUCT	lei	48360,0	24360,0	61880,0	26650,0	18750,0	20040,0	14080,0	214120,0	49795,3
D (-) TOTAL EXPENSES	lei	39947,9	21179,9	53732,0	22312,7	14597,6	18602,1	12165,4	182537,6	42450,6
D1. From the main production	lei	39947,4	21179,3	53730,7	22312,3	14597,4	18602,1	12165,2	182534,4	42449,9
I. VARIABLE COSTS	lei	35347,7	18669,1	42322,8	18597,7	12839,4	14817,0	10598,9	153192,5	35626,2
1. Raw materials and supplies expenditure	lei	15780,2	7637,4	22157,8	6219,0	4426,2	7966,2	4539,3	68726,1	15982,8
- Seedlings	lei	5850,0	2660,0	11135,0	1585,0	525,6	1207,8	1960,0	24923,4	5796,1
- Fertilizers	lei	5733,0	2598,4	5997,6	3607,0	3032,4	1113,6	660,4	22742,4	5288,9
- Pesticides	lei	3560,7	1834,7	5025,2	1027,0	609,0	5644,8	1757,6	19459,0	4525,3
- Other materials	lei	636,5	544,3	0,0	0,0	259,2	0,0	161,3	1601,3	372,4
2. Expenses with mechanical works	lei	18074,3	10276,6	18109,2	11678,0	7936,2	6120,7	5618,2	77813,1	18096,1

Indicators	U.M	CROP							TOTAL	
		Wheat	Barley	Corn	Sunflower	Rape	Sugar beet	Beans	lei/expl.	€/expl.
		ha								
3. Expenditure on irrigation	lei	0,0	0,0	0,0	0,0	0,0	0,0	0,0	0,0	0,0
4. Supply Expenses	lei	789,0	381,9	1107,9	311,0	221,3	398,3	227,0	3436,3	799,1
5. Temporary labor expenses	lei	0,0	0,0	0,0	0,0	0,0	0,0	0,0	0,0	0,0
6. Insurance	lei	704,2	373,2	947,9	389,8	255,7	331,8	214,4	3217,0	748,1
II. FIXED COSTS	lei	4600,2	2510,8	11409,3	3715,0	1758,2	3785,2	1566,5	29345,1	6824,4
- Permanent labor expenses	lei	568,5	364,3	6020,2	1279,8	200,9	2105,9	335,3	10874,8	2529,0
- Overheads	lei	1056,4	559,8	1421,9	584,6	383,5	497,7	321,6	4825,5	1122,2
- Interest	lei	2065,4	1096,7	2777,2	1150,6	753,8	971,6	629,6	9444,8	2196,5
- Depreciation	lei	910,0	490,0	1190,0	700,0	420,0	210,0	280,0	4200,0	976,7
E. (=) TAXABLE INCOME	lei	-947,4	-2139,3	-4770,7	-2812,3	-197,4	-602,1	-965,2	-12434,4	-2891,7
(-) Taxes	lei	0,0	0,0	0,0	0,0	0,0	0,0	0,0	0,0	0,0
F. (=) NET REVENUE+SUBSIDIES	lei	7892,6	2620,7	6789,3	3987,7	3882,6	1437,9	1754,8	28365,6	6596,7
G. RATA TAXABLE INCOME	%	-2,4	-10,1	-8,9	-12,6	-1,4	-3,2	-7,9	-6,8	-6,8
H. RATA NET REVENUE+SUBSIDIES	%	19,8	12,4	12,6	17,9	26,6	7,7	14,4	15,5	15,5

Source: author's own computation

Appendix 8b

BUDGET of 60 Ha Farm in Lowland Area
Option at the End of the Forecasted Range – Year 2017

Indicators	U.M	CROP							TOTAL	
		Wheat	Barley	Corn	Sunflower	Rape	Sugar beet	Beans	lei/expl.	€/expl.
		ha								
		13,00	7,00	17,00	10,00	6,00	3,00	4,00	60,00	60,00
		21,7%	11,7%	28,3%	16,7%	10,0%	5,0%	6,6%	100%	100%
A. PRODUCTION VALUE	lei	55120,0	27216,0	69904,0	27650,0	20430,0	25200,0	15840,0	241360,0	56130,2
A1. From the main production	lei	54600,0	26656,0	68544,0	27300,0	20160,0	25200,0	15680,0	238140,0	55381,4
B (+) SUBSIDIES	lei	8840,0	4760,0	11560,0	6800,0	4080,0	2040,0	2720,0	40800,0	9488,4
C (=) GROSS PRODUCT	lei	63960,0	31976,0	81464,0	34450,0	24510,0	27240,0	18560,0	282160,0	65618,6
D (-) TOTAL EXPENSES	lei	51132,9	27109,6	68775,2	28560,0	18684,6	23810,4	15571,2	233643,9	54335,8
D1. From the main production	lei	51132,4	27109,0	68773,8	28559,7	18684,3	23810,4	15571,0	233640,7	54335,0
I. VARIABLE COSTS	lei	45245,0	23896,4	54173,2	23805,0	16434,5	18965,7	13566,5	196086,4	45601,5
1. Raw materials and supplies expenditure	lei	20198,6	9775,9	28362,0	7960,3	5665,5	10196,7	5810,3	87969,4	20458,0
- Seedlings	lei	7488,0	3404,8	14252,8	2028,8	672,8	1546,0	2508,8	31902,0	7419,1
- Fertilizers	lei	7338,2	3326,0	7676,9	4617,0	3881,5	1425,4	845,3	29110,3	6769,8
- Pesticides	lei	4557,7	2348,4	6432,3	1314,6	779,5	7225,3	2249,7	24907,5	5792,4
- Other materials	lei	814,7	696,7	0,0	0,0	331,8	0,0	206,4	2049,6	476,7
2. Expenses with mechanical works	lei	23135,0	13154,1	23179,8	14947,8	10158,4	7834,4	7191,3	99600,8	23163,0
3. Expenditure on irrigation	lei	0,0	0,0	0,0	0,0	0,0	0,0	0,0	0,0	0,0

Indicators	U.M	CROP							TOTAL	
		Wheat	Barley	Corn	Sunflower	Rape	Sugar beet	Beans	lei/expl.	€/expl.
		ha								
4. Supply Expenses	lei	1009,9	488,8	1418,1	398,0	283,3	509,8	290,5	4398,5	1022,9
5. Temporary labor expenses	lei	0,0	0,0	0,0	0,0	0,0	0,0	0,0	0,0	0,0
6. Insurance	lei	901,4	477,7	1213,3	498,9	327,3	424,7	274,4	4117,8	957,6
II. FIXED COSTS	lei	5887,9	3213,2	14602,0	4755,0	2250,1	4844,7	2004,7	37557,5	8734,3
- Permanent labor expenses	lei	727,6	466,3	7705,8	1638,1	257,1	2695,5	429,2	13919,7	3237,1
- Overheads	lei	1352,1	716,6	1820,0	748,3	490,9	637,1	411,6	6176,7	1436,4
- Interest	lei	2643,7	1403,8	3554,9	1472,8	964,9	1243,6	805,8	12089,4	2811,5
- Depreciation	lei	1164,4	626,6	1521,4	895,8	537,2	268,5	357,9	5371,7	1249,2
E. (=) TAXABLE INCOME	lei	3467,6	-453,0	-229,8	-1259,7	1475,7	1389,6	109,0	4499,3	1046,4
(-) Taxes	lei	0,0	0,0	0,0	0,0	0,0	0,0	0,0	0,0	0,0
F. (=) NET REVENUE+SUBSIDIES	lei	12307,6	4307,0	11330,2	5540,4	5555,7	3429,6	2829,0	45299,3	10534,7
G. RATA TAXABLE INCOME	%	6,8	-1,7	-0,3	-4,4	7,9	5,8	0,7	1,9	1,9
H. RATA NET REVENUE+SUBSIDIES	%	24,1	15,9	16,5	19,4	29,7	14,4	18,2	19,4	19,4

Source: author's own computation

Appendix 9

Annual Budget for the Holding of 100 Ha in Lowland Area, 2013–2017

Indicators	U.M	2012		2013		2014		2015		2016		2017	
		lei/expl.	€/expl.	lei/expl.	€/expl.	lei/expl.	€/expl.	lei/expl.	€/expl.	lei/expl.	€/expl.	lei/expl.	€/expl.
A. PRODUCTION VALUE	lei	317543,0	73847,2	338409,0	78699,8	361451,0	84058,4	385877,0	89738,8	412837,0	96008,6	442347,0	102871,4
A[1]. From the main production	lei	312010,0	72560,5	332876,0	77413,0	355918,0	82771,6	380344,0	88452,1	407304,0	94721,9	436814,0	101584,7
B (+) SUBSIDIES	lei	68000,0	15814,0	68000,0	15814,0	68000,0	15814,0	68000,0	15814,0	68000,0	15814,0	68000,0	15814,0
C (=) GROSS PRODUCT	lei	385543,0	89661,2	406409,0	94513,7	429451,0	99872,3	453877,0	105552,8	480837,0	111822,6	510347,0	118685,3
D (-) TOTAL EXPENSES	lei	312219,9	72609,3	327825,7	76238,5	343437,1	79869,1	362170,5	84225,7	380903,9	88582,3	399636,2	92938,7
D[1]. From the main production	lei	312214,4	72608,0	327820,2	76237,2	343431,6	79867,8	362165,0	84224,4	380898,4	88581,0	399630,7	92937,4
I. VARIABLE COSTS	lei	264106,2	61420,0	277311,5	64491,1	290516,8	67562,1	306363,2	71247,3	322209,6	74932,5	338055,9	78617,7
1. Raw materials and supplies expenditure	lei	116861,5	27177,1	122704,6	28535,9	128547,7	29894,8	135559,3	31525,4	142571,0	33156,1	149582,7	34786,7
- Seedlings	lei	39970,2	9295,4	41968,7	9760,2	43967,2	10224,9	46365,4	10782,7	48763,6	11340,4	51161,9	11898,1
- Fertilizers	lei	40639,6	9451,1	42671,6	9923,6	44703,6	10396,2	47141,9	10963,2	49580,3	11530,3	52018,7	12097,4
- Pesticides	lei	32954,1	7663,7	34601,8	8046,9	36249,5	8430,1	38226,8	8889,9	40204,0	9349,8	42181,2	9809,6
- Other materials	lei	3297,6	766,9	3462,5	805,2	3627,4	843,6	3825,2	889,6	4023,1	935,6	4220,9	981,6
2. Expenses with mechanical works	lei	135895,9	31603,7	142690,7	33183,9	149485,4	34764,1	157639,2	36660,3	165793,0	38556,5	173946,7	40452,7
3. Expenditure on irrigation	lei	0,0	0,0	0,0	0,0	0,0	0,0	0,0	0,0	0,0	0,0	0,0	0,0

Indicators	U.M	2012		2013		2014		2015		2016		2017	
		lei/expl.	€/expl.	lei/expl.	€/expl.	lei/expl.	€/expl.	lei/expl.	€/expl.	lei/expl.	€/expl.	lei/expl.	€/expl.
4. Supply Expenses	lei	5843,1	1358,9	6135,2	1426,8	6427,4	1494,7	6778,0	1576,3	7128,6	1657,8	7479,1	1739,3
5. Temporary labor expenses	lei	0,0	0,0	0,0	0,0	0,0	0,0	0,0	0,0	0,0	0,0	0,0	0,0
6. Insurance	lei	5505,8	1280,4	5781,1	1344,4	6056,4	1408,5	6386,7	1485,3	6717,0	1562,1	7047,4	1638,9
II. FIXED COSTS	lei	48113,7	11189,2	50514,2	11747,5	52920,3	12307,0	55807,3	12978,4	58694,3	13649,8	61580,3	14321,0
- Permanent labor expenses	lei	16688,3	3881,0	17522,7	4075,1	18357,1	4269,1	19358,4	4502,0	20359,7	4734,8	21361,0	4967,7
- Overheads	lei	8258,7	1920,6	8671,6	2016,7	9084,5	2112,7	9580,0	2227,9	10075,6	2343,2	10571,1	2458,4
- Interest	lei	16166,7	3759,7	16975,1	3947,7	17783,4	4135,7	18753,4	4361,3	19723,4	4586,8	20693,4	4812,4
- Depreciation	lei	7000,0	1627,9	7344,8	1708,1	7695,2	1789,6	8115,4	1887,3	8535,6	1985,0	8954,7	2082,5
E. (=) TAXABLE INCOME	lei	-204,4	-47,5	5055,8	1175,8	12486,4	2903,8	18179,0	4227,7	26405,6	6140,8	37183,3	8647,3
(-) Taxes	lei	0,0	0,0	0,0	0,0	0,0	0,0	0,0	0,0	0,0	0,0	0,0	0,0
F. (=) NET REVENUE+ SUBSIDIES	lei	67795,6	15766,4	73055,8	16989,7	80486,4	18717,8	86179,0	20041,6	94405,6	21954,8	105183,3	24461,2
G. RATA TAXABLE INCOME	%	-0,1	-0,1	1,5	1,5	3,6	3,6	5,0	5,0	6,9	6,9	9,3	9,3
H. RATA NET REVENUE+ SUBSIDIES	%	21,7	21,7	22,3	22,3	23,4	23,4	23,8	23,8	24,8	24,8	26,3	26,3

Source: author's own computation

Appendix 9a

BUDGET of 100 Ha Farm in Lowland Area
Basic variant – year 2012

Indicators	U.M	CROP							TOTAL	
		Wheat	Barley	Corn	Sunflower	Beans	Sugar beet	Rape	lei/expl.	€/expl.
		ha								
		22,00	11,00	22,00	16,00	12,00	5,00	12,00	100,00	100,00
		22,0%	11,0%	22,0%	16,0%	12,0%	5,0%	12,0%	100%	100%
A. PRODUCTION VALUE	lei	71368,0	32725,0	69190,0	34160,0	42600,0	34500,0	33000,0	317543,0	73847,2
A^1. From the main production	lei	70400,0	31790,0	67320,0	33600,0	42000,0	34500,0	32400,0	312010,0	72560,5
B (+) SUBSIDIES	lei	14960,0	7480,0	14960,0	10880,0	8160,0	3400,0	8160,0	68000,0	15814,0
C (=) GROSS PRODUCT	lei	86328,0	40205,0	84150,0	45040,0	50760,0	37900,0	41160,0	385543,0	89661,2
D (-) TOTAL EXPENSES	lei	69606,1	34175,2	70531,2	36474,8	38753,6	31651,3	31027,6	312219,9	72609,3
D^1. From the main production	lei	69605,2	34174,3	70529,3	36474,3	38753,0	31651,3	31027,0	312214,4	72608,0
I. VARIABLE COSTS	lei	61601,3	30132,6	55528,0	30469,1	33816,3	25225,0	27333,9	264106,2	61420,0
1. Raw materials and supplies expenditure	lei	27515,8	12320,2	29185,2	10366,4	14193,1	13560,0	9720,7	116861,5	27177,1
- Seedlings	lei	9900,0	4180,0	14410,0	2536,0	5880,0	2013,0	1051,2	39970,2	9295,4
- Fertilizers	lei	10386,2	4338,4	8272,0	6187,2	2452,8	2139,0	6864,0	40639,6	9451,1
- Pesticides	lei	6025,8	2883,1	6503,2	1643,2	5272,8	9408,0	1218,0	32954,1	7663,7
- Other materials	lei	1203,8	918,7	0,0	0,0	587,5	0,0	587,5	3297,6	766,9
2. Expenses with mechanical works	lei	31481,7	16593,8	23638,9	18946,8	18229,7	10422,2	16582,7	135895,9	31603,7
3. Expenditure on irrigation	lei	0,0	0,0	0,0	0,0	0,0	0,0	0,0	0,0	0,0

Indicators	U.M	CROP							TOTAL	
		Wheat	Barley	Corn	Sunflower	Beans	Sugar beet	Rape	lei/expl.	€/expl.
		ha								
4. Supply Expenses	lei	1375,8	616,0	1459,3	518,3	709,7	678,0	486,0	5843,1	1358,9
5. Temporary labor expenses	lei	0,0	0,0	0,0	0,0	0,0	0,0	0,0	0,0	0,0
6. Insurance	lei	1227,9	602,6	1244,7	637,6	683,9	564,7	544,4	5505,8	1280,4
II. FIXED COSTS	lei	8004,9	4042,6	15003,1	6005,7	4937,3	6426,4	3693,7	48113,7	11189,2
- Permanent labor expenses	lei	1021,9	598,1	7949,7	2047,7	1062,4	3575,8	432,8	16688,3	3881,0
- Overheads	lei	1841,9	903,8	1867,0	956,4	1025,8	847,1	816,7	8258,7	1920,6
- Interest	lei	3601,2	1770,7	3646,4	1881,7	2009,1	1653,5	1604,2	16166,7	3759,7
- Depreciation	lei	1540,0	770,0	1540,0	1120,0	840,0	350,0	840,0	7000,0	1627,9
E. (=) TAXABLE INCOME	lei	794,8	-2384,3	-3209,3	-2874,3	3247,0	2848,7	1373,0	-204,4	-47,5
(-) Taxes	lei	0,0	0,0	0,0	0,0	0,0	0,0	0,0	0,0	0,0
F. (=) NET REVENUE+SUBSIDIES	lei	15754,8	5095,7	11750,7	8005,7	11407,0	6248,7	9533,0	67795,6	15766,4
G. RATA TAXABLE INCOME	%	1,1	-7,0	-4,6	-7,9	8,4	9,0	4,4	-0,1	-0,1
H. RATA NET REVENUE+SUBSIDIES	%	22,6	14,9	16,7	21,9	29,4	19,7	30,7	21,7	21,7

Source: author's own computation

Appendix 9b

Budget for the holding of 100 ha in Lowland area
Option at the End of the Forecasted Range – Year 2017

Indicators	U.M	CROP							TOTAL	
		Wheat	Barley	Corn	Sunflower	Beans	Sugar beet	Rape	lei/expl.	€/expl.
		ha								
		22,00	11,00	22,00	16,00	12,00	5,00	12,00	100,00	100,00
		22,0%	11,0%	22,0%	16,0%	12,0%	5,0%	12,0%	100%	100%
A. PRODUCTION VALUE	lei	99528,0	45441,0	96118,0	47600,0	59400,0	48300,0	45960,0	442347,0	102871,4
A1. From the main production	lei	98560,0	44506,0	94248,0	47040,0	58800,0	48300,0	45360,0	436814,0	101584,7
B (+) SUBSIDIES	lei	14960,0	7480,0	14960,0	10880,0	8160,0	3400,0	8160,0	68000,0	15814,0
C (=) GROSS PRODUCT	lei	114488,0	52921,0	111078,0	58480,0	67560,0	51700,0	54120,0	510347,0	118685,3
D (-) TOTAL EXPENSES	lei	89093,4	43743,7	90279,2	46688,0	49604,4	40513,5	39714,0	399636,2	92938,7
D1. From the main production	lei	89092,4	43742,8	90277,3	46687,4	49603,8	40513,5	39713,4	399630,7	92937,4
I. VARIABLE COSTS	lei	78849,6	38569,7	71075,9	39000,5	43284,9	32288,0	34987,4	338055,9	78617,7
1. Raw materials and supplies expenditure	lei	35220,3	15769,9	37357,1	13269,0	18167,2	17356,8	12442,5	149582,7	34786,7
- Seedlingsr	lei	12672,0	5350,4	18444,8	3246,1	7526,4	2576,6	1345,5	51161,9	11898,1
- Fertilizers	lei	13294,3	5553,2	10588,2	7919,6	3139,6	2737,9	8785,9	52018,7	12097,4
- Pesticides	lei	7713,0	3690,4	8324,1	2103,3	6749,2	12042,2	1559,0	42181,2	9809,6
- Other materials	lei	1540,9	1176,0	0,0	0,0	752,0	0,0	752,0	4220,9	981,6
2. Expenses with mechanical works	lei	40296,6	21240,1	30257,8	24251,9	23334,0	13340,5	21225,9	173946,7	40452,7
3. Expenditure on irrigation	lei	0,0	0,0	0,0	0,0	0,0	0,0	0,0	0,0	0,0
4. Supply Expenses	lei	1761,0	788,5	1867,9	663,4	908,4	867,8	622,1	7479,1	1739,3
5. Temporary labor expenses	lei	0,0	0,0	0,0	0,0	0,0	0,0	0,0	0,0	0,0

Indicators	U.M	CROP							TOTAL	
		Wheat	Barley	Corn	Sunflower	Beans	Sugar beet	Rape	lei/expl.	€/expl.
		ha								
6. Insurance	lei	1571,7	771,3	1593,2	816,1	875,4	722,8	696,9	7047,4	1638,9
II. FIXED COSTS	lei	10243,8	5174,0	19203,3	7687,5	6319,5	8225,5	4726,6	61580,3	14321,0
- Permanent labor expenses	lei	1308,0	765,5	10175,6	2621,0	1359,8	4577,0	554,0	21361,0	4967,7
- Overheads	lei	2357,6	1156,9	2389,8	1224,2	1313,1	1084,3	1045,3	10571,1	2458,4
- Interest	lei	4609,5	2266,5	4667,4	2408,5	2571,7	2116,5	2053,3	20693,4	4812,4
- Depreciation	lei	1968,7	985,0	1970,5	1433,8	1075,0	447,8	1073,9	8954,7	2082,5
E. (=) TAXABLE INCOME	lei	9467,6	763,2	3970,7	352,6	9196,2	7786,5	5646,6	37183,3	8647,3
(-) Taxes	lei	0,0	0,0	0,0	0,0	0,0	0,0	0,0	0,0	0,0
F. (=) NET REVENUE+SUBSIDIES	lei	24427,6	8243,2	18930,7	11232,6	17356,2	11186,5	13806,6	105183,3	24461,2
G. RATA TAXABLE INCOME	%	10,6	1,7	4,4	0,8	18,5	19,2	14,2	9,3	9,3
H. RATA NET REVENUE+SUBSIDIES	%	27,4	18,8	21,0	24,1	35,0	27,6	34,8	26,3	26,3

Source: author's own computation

Appendix 10

Annual Budget for the Holding of 250 Ha in Lowland Area, 2013–2017

Indicators	U.M	2012		2013		2014		2015		2016		2017	
		lei/expl.	€/expl.	lei/expl.	€/expl.	lei/expl.	€/expl.	lei/expl.	€/expl.	lei/expl.	€/expl.	lei/expl.	€/expl.
A. PRODUCTION VALUE	lei	925890,0	215323,3	986465,0	229410,5	1054780,0	245297,7	1127260,0	262153,5	1203905,0	279977,9	1289090,0	299788,4
A1. From the main production	lei	908000,0	211162,8	968575,0	225250,0	1036890,0	241137,2	1109370,0	257993,0	1186015,0	275817,4	1271200,0	295627,9
B (+) SUBSIDIES	lei	170000,0	39534,9	170000,0	39534,9	170000,0	39534,9	170000,0	39534,9	170000,0	39534,9	170000,0	39534,9
C (=) GROSS PRODUCT	lei	1095890,0	254858,1	1156465,0	268945,3	1224780,0	284832,6	1297260,0	301688,4	1373905,0	319512,8	1459090,0	339323,3
D (-) TOTAL EXPENSES	lei	885377,2	205901,7	929641,0	216195,6	973910,5	226490,8	1027028,0	238843,7	1080150,0	251197,7	1133273,0	263551,9
D1. From the main production	lei	885359,3	205897,5	929623,1	216191,4	973892,6	226486,7	1027010,1	238839,6	1080132,1	251193,5	1133255,1	263547,7
I. VARIABLE COSTS	lei	737892,7	171603,0	774787,4	180183,1	811682,0	188763,3	855955,6	199059,4	900229,1	209355,6	944502,7	219651,8
1. Raw materials and supplies expenditure	lei	294698,6	68534,6	309433,5	71961,3	324168,5	75388,0	341850,4	79500,1	359532,3	83612,2	377214,2	87724,2
- Seedlings	lei	99618,0	23167,0	104598,9	24325,3	109579,8	25483,7	115556,9	26873,7	121534,0	28263,7	127511,0	29653,7
- Fertilizers	lei	112239,0	26102,1	117851,0	27407,2	123462,9	28712,3	130197,2	30278,4	136931,6	31844,6	143665,9	33410,7
- Pesticides	lei	72920,0	16958,1	76566,0	17806,0	80212,0	18654,0	84587,2	19671,4	88962,4	20688,9	93337,6	21706,4
- Other materials	lei	9921,6	2307,3	10417,7	2422,7	10913,8	2538,1	11509,1	2676,5	12104,4	2815,0	12699,6	2953,4
2. Expenses with mechanical works	lei	335859,9	78107,0	352652,9	82012,3	369445,9	85917,6	389597,5	90604,1	409749,1	95290,5	429900,7	99976,9

Indicators	U.M	2012		2013		2014		2015		2016		2017	
		lei/expl.	€/expl.	lei/expl.	€/expl.	lei/expl.	€/expl.	lei/expl.	€/expl.	lei/expl.	€/expl.	lei/expl.	€/expl.
3. Expenditure on irrigation	lei	77000,0	17907,0	80850,0	18802,3	84700,0	19697,7	89320,0	20772,1	93940,0	21846,5	98560,0	22920,9
4. Supply Expenses	lei	14734,9	3426,7	15471,7	3598,1	16208,4	3769,4	17092,5	3975,0	17976,6	4180,6	18860,7	4386,2
5. Temporary labor expenses	lei	0,0	0,0	0,0	0,0	0,0	0,0	0,0	0,0	0,0	0,0	0,0	0,0
6. Insurance	lei	15599,3	3627,7	16379,3	3809,1	17159,2	3990,5	18095,2	4208,2	19031,1	4425,8	19967,1	4643,5
II. FIXED COSTS	lei	147484,5	34298,7	154853,6	36012,5	162228,5	37727,6	171072,4	39784,3	179920,9	41842,1	188770,3	43900,1
- Permanent labor expenses	lei	57671,7	13412,0	60555,3	14082,6	63438,9	14753,2	66899,2	15557,9	70359,5	16362,7	73819,8	17167,4
- Overheads	lei	23399,0	5441,6	24568,9	5713,7	25738,8	5985,8	27142,8	6312,3	28546,7	6638,8	29950,7	6965,3
- Interest	lei	45913,8	10677,6	48209,5	11211,5	50505,2	11745,4	53260,0	12386,1	56014,8	13026,7	58769,7	13667,4
- Depreciation	lei	20500,0	4767,4	21520,0	5004,6	22545,6	5243,2	23770,5	5528,0	24999,8	5813,9	26230,2	6100,0
E. (=) TAXABLE INCOME	lei	22640,7	5265,3	38951,9	9058,6	62997,4	14650,6	82359,9	19153,5	105882,9	24623,9	137944,9	32080,2
(-) Taxes	lei	0,0	0,0	0,0	0,0	0,0	0,0	0,0	0,0	0,0	0,0	0,0	0,0
F. (=) NET REVENUE + SUBSIDIES	lei	192640,7	44800,2	208951,9	48593,5	232997,4	54185,4	252359,9	58688,3	275882,9	64158,8	307944,9	71615,1
G. RATA TAXABLE INCOME	%	2,6	2,6	4,2	4,2	6,5	6,5	8,0	8,0	9,8	9,8	12,2	12,2
H. RATA NET REVENUE + SUBSIDIES	%	21,8	21,8	22,5	22,5	23,9	23,9	24,6	24,6	25,5	25,5	27,2	27,2

Source: author's own computation

Appendix 10a

Budget of 250 Ha Holding in Lowland Area
Basic Variant – Year 2012

Indicators	U.M	CROP						TOTAL	
		Wheat	Barley	Corn	Sunflower	Soy	Rape	lei/expl.	€/expl.
		ha							
		55,00	25,00	50,00	40,00	50,00	30,00	250,00	250,00
		22,0%	10,0%	20,0%	16,0%	20,0%	12,0%	100%	100%
A. PRODUCTION VALUE	lei	195140,0	78750,0	249750,0	97600,0	213000,0	91650,0	925890,0	215323,3
A1. From the main production	lei	192500,0	76500,0	243000,0	96000,0	210000,0	90000,0	908000,0	211162,8
B (+) SUBSIDIES	lei	37400,0	17000,0	34000,0	27200,0	34000,0	20400,0	170000,0	39534,9
C (=) GROSS PRODUCT	lei	232540,0	95750,0	283750,0	124800,0	247000,0	112050,0	1095890,0	254858,1
D (-) TOTAL EXPENSES	lei	180317,4	79700,3	238763,8	88290,3	216220,2	82085,1	885377,2	205901,7
D1. From the main production	lei	180314,8	79698,1	238757,0	88288,7	216217,2	82083,4	885359,3	205897,5
I. VARIABLE COSTS	lei	158539,1	70291,0	181377,7	78504,0	176768,1	72412,8	737892,7	171603,0
1. Raw materials and supplies expenditure	lei	70042,5	28724,5	73225,0	26956,0	69335,0	26415,6	294698,6	68534,6
- Seedlings	lei	24750,0	9500,0	32750,0	6340,0	23650,0	2628,0	99618,0	23167,0
- Fertilizers	lei	27060,0	10440,0	25695,0	16508,0	13435,0	19101,0	112239,0	26102,1
- Pesticides	lei	15064,5	6552,5	14780,0	4108,0	29370,0	3045,0	72920,0	16958,1
- Other materials	lei	3168,0	2232,0	0,0	0,0	2880,0	1641,6	9921,6	2307,3
2. Expenses with mechanical works	lei	81811,1	38724,2	58275,5	48658,4	65156,8	43233,9	335859,9	78107,0
3. Expenditure on irrigation	lei	0,0	0,0	42000,0	0,0	35000,0	0,0	77000,0	17907,0
4. Supply Expenses	lei	3502,1	1436,2	3661,3	1347,8	3466,8	1320,8	14734,9	3426,7
5. Temporary labor expenses	lei	0,0	0,0	0,0	0,0	0,0	0,0	0,0	0,0

Indicators	U.M	CROP						TOTAL	
		Wheat	Barley	Corn	Sunflower	Soy	Rape	lei/expl.	€/expl.
		ha							
6. Insurance	lei	3183,4	1406,1	4215,9	1541,8	3809,5	1442,6	15599,3	3627,7
II. FIXED COSTS	lei	21778,3	9409,4	57386,1	9786,3	39452,1	9672,2	147484,5	34298,7
- Permanent labor expenses	lei	3813,2	1418,3	33634,2	129,0	17517,3	1159,8	57671,7	13412,0
- Overheads	lei	4775,1	2109,1	6323,9	2312,7	5714,3	2163,9	23399,0	5441,6
- Interest	lei	9340,0	4132,0	12428,1	4544,6	11220,5	4248,6	45913,8	10677,6
- Depreciation	lei	3850,0	1750,0	5000,0	2800,0	5000,0	2100,0	20500,0	4767,4
E. (=) TAXABLE INCOME	lei	12185,2	-3198,1	4243,0	7711,3	-6217,2	7916,6	22640,7	5265,3
(-) Taxes	lei	0,0	0,0	0,0	0,0	0,0	0,0	0,0	0,0
F. (=) NET REVENUE+SUBSIDIES	lei	49585,2	13801,9	38243,0	34911,3	27782,8	28316,6	192640,7	44800,2
G. RATA TAXABLE INCOME	%	6,8	-4,0	1,8	8,7	-2,9	9,6	2,6	2,6
H. RATA NET REVENUE+SUBSIDIES	%	27,5	17,3	16,0	39,5	12,8	34,5	21,8	21,8

Source: author's own computation

Appendix 10b

Budget of 250 Ha Holding in Lowland Area
Projected Variant – Year 2017

Indicators	U.M	CROP						TOTAL	
		Wheat	Barley	Corn	Sunflower	Soy	Rape	lei/expl.	€/expl.
		ha							
		55,00	25,00	50,00	40,00	50,00	30,00	250,00	250,00
		22,0%	10,0%	20,0%	16,0%	20,0%	12,0%	100%	100%
A. PRODUCTION VALUE	lei	272140,0	109350,0	346950,0	136000,0	297000,0	127650,0	1289090,0	299788,4
A1. From the main production	lei	269500,0	107100,0	340200,0	134400,0	294000,0	126000,0	1271200,0	295627,9
B (+) SUBSIDIES	lei	37400,0	17000,0	34000,0	27200,0	34000,0	20400,0	170000,0	39534,9
C (=) GROSS PRODUCT	lei	309540,0	126350,0	380950,0	163200,0	331000,0	148050,0	1459090,0	339323,3
D (-) TOTAL EXPENSES	lei	230802,0	102015,0	305615,0	113012,0	276760,0	105069,0	1133273,0	263551,9
D1. From the main production	lei	230799,4	102012,8	305608,3	113010,4	276757,0	105067,4	1133255,1	263547,7
I. VARIABLE COSTS	lei	202930,1	89972,4	232163,4	100485,1	226263,2	92688,4	944502,7	219651,8
1. Raw materials and supplies expenditure	lei	89654,4	36767,4	93728,0	34503,7	88748,8	33812,0	377214,2	87724,2
- Seedlings	lei	31680,0	12160,0	41920,0	8115,2	30272,0	3363,8	127511,0	29653,7
- Fertilizers	lei	34636,8	13363,2	32889,6	21130,2	17196,8	24449,3	143665,9	33410,7
- Pesticides	lei	19282,6	8387,2	18918,4	5258,2	37593,6	3897,6	93337,6	21706,4
- Other materials	lei	4055,0	2857,0	0,0	0,0	3686,4	2101,2	12699,6	2953,4
2. Expenses with mechanical works	lei	104718,3	49566,9	74592,6	62282,7	83400,7	55339,4	429900,7	99976,9
3. Expenditure on irrigation	lei	0,0	0,0	53760,0	0,0	44800,0	0,0	98560,0	22920,9

Indicators	U.M	CROP						TOTAL	
		Wheat	Barley	Corn	Sunflower	Soy	Rape	lei/expl.	€/expl.
		ha							
4. Supply Expenses	lei	4482,7	1838,4	4686,4	1725,2	4437,4	1690,6	18860,7	4386,2
5. Temporary labor expenses	lei	0,0	0,0	0,0	0,0	0,0	0,0	0,0	0,0
6. Insurance	lei	4074,7	1799,8	5396,4	1973,5	4876,2	1846,5	19967,1	4643,5
II. FIXED COSTS	lei	27871,9	12042,6	73451,6	12526,9	50496,8	12380,6	188770,3	43900,1
- Permanent labor expenses	lei	4880,9	1815,4	43051,7	165,1	22422,2	1484,5	73819,8	17167,4
- Overheads	lei	6112,1	2699,6	8094,6	2960,3	7314,3	2769,8	29950,7	6965,3
- Interest	lei	11955,2	5289,0	15907,9	5817,1	14362,3	5438,1	58769,7	13667,4
- Depreciation	lei	4923,7	2238,6	6397,4	3584,4	6398,1	2688,1	26230,2	6100,0
E. (=) TAXABLE INCOME	lei	38700,6	5087,3	34591,8	21389,6	17243,0	20932,7	137944,9	32080,2
(-) Taxes	lei	0,0	0,0	0,0	0,0	0,0	0,0	0,0	0,0
F. (=) NET REVENUE+SUBSIDIES	lei	76100,6	22087,3	68591,8	48589,6	51243,0	41332,7	307944,9	71615,1
G. RATA TAXABLE INCOME (%)	%	16,8	5,0	11,3	18,9	6,2	19,9	12,2	12,2
H. RATA NET REVENUE+SUBSIDIES	%	33,0	21,7	22,4	43,0	18,5	39,3	27,2	27,2

Source: author's own computation

Appendix 11

Annual Budget For The Holding of 500 Ha in Lowland Area, 2013–2017

Indicators	U.M	2012		2013		2014		2015		2016		2017	
		lei/expl.	€/expl.	lei/expl.	€/expl.	lei/expl.	€/expl.	lei/expl.	€/expl.	lei/expl.	€/expl.	lei/expl.	€/expl.
A. PRODUCTION VALUE	lei	2038500,0	474069,8	2171700,0	505046,5	2322280,0	540065,1	2481840,0	577172,1	2650380,0	616367,4	2838100,0	660023,3
A1. From the main production	lei	1999000,0	464883,7	2132200,0	495860,5	2282780,0	530879,1	2442340,0	567986,0	2610880,0	607181,4	2798600,0	650837,2
B (+) SUBSIDIES	lei	340000,0	79069,8	340000,0	79069,8	340000,0	79069,8	340000,0	79069,8	340000,0	79069,8	340000,0	79069,8
C (=) GROSS PRODUCT	lei	2378500,0	553139,5	2511700,0	584116,3	2662280,0	619134,9	2821840,0	656241,9	2990380,0	695437,2	3178100,0	739093,0
D (-) TOTAL EXPENSES	lei	1834454,6	426617,3	1926163,0	447944,9	2017887,0	469276,0	2129512,0	495235,3	2238010,0	520467,4	2348080,0	546065,1
D1. From the main production	lei	1834415,1	426608,2	1926123,5	447935,7	2017847,5	469266,9	2129472,5	495226,2	2237970,5	520458,3	2348040,5	546055,9
I. VARIABLE COSTS	lei	1523431,0	354286,3	1599602,6	372000,6	1675774,1	389714,9	1767180,0	410972,1	1858585,8	432229,3	1949991,7	453486,4
1. Raw materials and supplies expenditure	lei	610605,0	142001,2	641135,3	149101,2	671665,5	156201,3	708301,8	164721,3	744938,1	173241,4	781574,4	181761,5
- Seedlings	lei	195612,0	45491,2	205392,6	47765,7	215173,2	50040,3	226909,9	52769,7	238646,6	55499,2	250383,4	58228,7
- Fertilizers	lei	249277,0	57971,4	261740,9	60870,0	274204,7	63768,5	289161,3	67246,8	304117,9	70725,1	319074,6	74203,4
- Pesticides	lei	144116,0	33515,3	151321,8	35191,1	158527,6	36866,9	167174,6	38877,8	175821,5	40888,7	184468,5	42899,6
- Other materials	lei	21600,0	5023,3	22680,0	5274,4	23760,0	5525,6	25056,0	5827,0	26352,0	6128,4	27648,0	6429,8
2. Expenses with mechanical works	lei	695948,2	161848,4	730745,7	169940,9	765543,1	178033,3	807300,0	187744,2	849056,9	197455,1	890813,8	207166,0

Indicators	U.M	2012		2013		2014		2015		2016		2017	
		lei/expl.	€/expl.	lei/expl.	€/expl.	lei/expl.	€/expl.	lei/expl.	€/expl.	lei/expl.	€/expl.	lei/expl.	€/expl.
3. Expenditure on irrigation	lei	154000,0	35814,0	161700,0	37604,7	169400,0	39395,3	178640,0	41544,2	187880,0	43693,0	197120,0	45841,9
4. Supply Expenses	lei	30530,3	7100,1	32056,8	7455,1	33583,3	7810,1	35415,1	8236,1	37246,9	8662,1	39078,7	9088,1
5. Temporary labor expenses	lei	0,0	0,0	0,0	0,0	0,0	0,0	0,0	0,0	0,0	0,0	0,0	0,0
6. Insurance	lei	32347,5	7522,7	33964,9	7898,8	35582,3	8274,9	37523,1	8726,3	39464,0	9177,7	41404,8	9629,0
II. FIXED COSTS	lei	311023,6	72331,1	326560,4	75944,3	342112,9	79561,1	362332,0	84263,3	379424,2	88238,2	398088,3	92578,7
- Permanent labor expenses	lei	126291,6	29370,1	132606,2	30838,7	138920,8	32307,2	146498,3	34069,4	154075,8	35831,6	161653,3	37593,8
- Overheads	lei	48521,3	11284,0	50947,3	11848,2	53373,4	12412,4	56284,7	13089,5	59195,9	13766,5	62107,2	14443,5
- Interest	lei	95210,7	22142,0	99971,2	23249,1	104731,8	24356,2	110444,4	25684,7	116157,0	27013,3	121869,7	28341,8
- Depreciation	lei	41000,0	9534,9	43035,7	10008,3	45087,0	10485,3	49104,7	11419,7	49995,4	11626,8	52458,2	12199,6
E. (=) TAXABLE INCOME	lei	164584,9	38275,6	206076,5	47924,8	264932,5	61612,2	312867,5	72759,9	372909,5	86723,1	450559,5	104781,3
(-) Taxes	lei	0,0	0,0	0,0	0,0	0,0	0,0	0,0	0,0	0,0	0,0	0,0	0,0
F. (=) NET REVENUE + SUBSIDIES	lei	504584,9	117345,3	546076,5	126994,5	604932,5	140682,0	652867,5	151829,7	712909,5	165792,9	790559,5	183851,0
G. RATA TAXABLE INCOME (%)	%	9,0	9,0	10,7	10,7	13,1	13,1	14,7	14,7	16,7	16,7	19,2	19,2
H. RATA NET REVENUE + SUBSIDIES	%	27,5	27,5	28,4	28,4	30,0	30,0	30,7	30,7	31,9	31,9	33,7	33,7

Source: author's own computation

Appendix 11a

Budget of 500 Ha Holding in Lowland Area
Basic Variant–Year 2012

Indicators	U.M	CROP						TOTAL	
		Wheat	Barley	Corn	Sunflower	Soy	Rape	lei/expl.	€/expl.
		ha							
		100,00	50,00	100,00	80,00	100,00	70,00	500,00	500,00
		20,0%	10,0%	20,0%	16,0%	20,0%	14,0%	100%	100%
A. PRODUCTION VALUE	lei	385200,0	175000,0	555000,0	219600,0	468500,0	235200,0	2038500,0	474069,8
A1. From the main production	lei	380000,0	170000,0	540000,0	216000,0	462000,0	231000,0	1999000,0	464883,7
B (+) SUBSIDIES	lei	68000,0	34000,0	68000,0	54400,0	68000,0	47600,0	340000,0	79069,8
C (=) GROSS PRODUCT	lei	453200,0	209000,0	623000,0	274000,0	536500,0	282800,0	2378500,0	553139,5
D (-) TOTAL EXPENSES	lei	337485,6	167348,4	489609,8	195733,4	442710,1	201567,4	1834454,6	426617,3
D1. From the main production	lei	337480,4	167343,4	489594,8	195729,8	442703,6	201563,2	1834415,1	426608,2
I. VARIABLE COSTS	lei	298860,5	147658,2	371709,9	164532,7	362655,5	178014,2	1523431,0	354286,3
1. Raw materials and supplies expenditure	lei	131876,0	60201,0	152160,0	58256,0	141976,0	66136,0	610605,0	142001,2
- Seedlingsr	lei	45000,0	19000,0	65500,0	12680,0	47300,0	6132,0	195612,0	45491,2
- Fertilizers	lei	53150,0	23200,0	57100,0	37360,0	29600,0	48867,0	249277,0	57971,4
- Pesticides	lei	27390,0	13105,0	29560,0	8216,0	58740,0	7105,0	144116,0	33515,3
- Other materials	lei	6336,0	4896,0	0,0	0,0	6336,0	4032,0	21600,0	5023,3
2. Expenses with mechanical works	lei	154428,9	81491,7	119292,0	99934,9	135776,5	105024,3	695948,2	161848,4
3. Expenditure on irrigation	lei	0,0	0,0	84000,0	0,0	70000,0	0,0	154000,0	35814,0

Indicators	U.M	CROP						TOTAL	
		Wheat	Barley	Corn	Sunflower	Soy	Rape	lei/expl.	€/expl.
		ha							
4. Supply Expenses	lei	6593,8	3010,1	7608,0	2912,8	7098,8	3306,8	30530,3	7100,1
5. Temporary labor expenses	lei	0,0	0,0	0,0	0,0	0,0	0,0	0,0	0,0
6. Insurance	lei	5961,8	2955,5	8649,8	3429,0	7804,3	3547,1	32347,5	7522,7
II. FIXED COSTS	lei	38625,1	19690,2	117899,9	31200,7	80054,5	23553,2	311023,6	72331,1
- Permanent labor expenses	lei	5192,5	3071,1	69432,1	10345,0	35361,3	2889,5	126291,6	29370,1
- Overheads	lei	8942,7	4433,2	12974,8	5143,5	11706,4	5320,7	48521,3	11284,0
- Interest	lei	17489,8	8685,8	25493,0	10112,2	22986,8	10443,0	95210,7	22142,0
- Depreciation	lei	7000,0	3500,0	10000,0	5600,0	10000,0	4900,0	41000,0	9534,9
E. (=) TAXABLE INCOME	lei	42519,6	2656,6	50405,2	20270,2	19296,4	29436,8	164584,9	38275,6
(-) Taxes	lei	0,0	0,0	0,0	0,0	0,0	0,0	0,0	0,0
F. (=) NET REVENUE+SUBSIDIES	lei	110519,6	36656,6	118405,2	74670,2	87296,4	77036,8	504584,9	117345,3
G. RATA TAXABLE INCOME (%)	%	12,6	1,6	10,3	10,4	4,4	14,6	9,0	9,0
H. RATA NET REVENUE+SUBSIDIES (%)	%	32,7	21,9	24,2	38,1	19,7	38,2	27,5	27,5

Source: author's own computation

Appendix 11b

Budget of 500 Ha Holding in Lowland Area
Option at the End of the Forecasted Range–Year 2017

Indicators	U.M	CROP						TOTAL	
		Wheat	Barley	Corn	Sunflower	Soy	Rape	lei/expl.	€/expl.
		ha							
		100,00	50,00	100,00	80,00	100,00	70,00	500,00	500,00
		20,0%	10,0%	20,0%	16,0%	20,0%	14,0%	100%	100%
A. PRODUCTION VALUE	lei	537200,0	243000,0	771000,0	306000,0	653300,0	327600,0	2838100,0	660023,3
A1. From the main production	lei	532000,0	238000,0	756000,0	302400,0	646800,0	323400,0	2798600,0	650837,2
B (+) SUBSIDIES	lei	68000,0	34000,0	68000,0	54400,0	68000,0	47600,0	340000,0	79069,8
C (=) GROSS PRODUCT	lei	605200,0	277000,0	839000,0	360400,0	721300,0	375200,0	3178100,0	739093,0
D (-) TOTAL EXPENSES	lei	431980,0	214205,0	626700,0	250536,0	566660,0	257999,0	2348080,0	546065,1
D1. From the main production	lei	431974,8	214200,0	626685,0	250532,4	566653,5	257994,8	2348040,5	546055,9
I. VARIABLE COSTS	lei	382541,4	189002,5	475788,6	210601,9	464199,1	227858,2	1949991,7	453486,4
1. Raw materials and supplies expenditure	lei	168801,3	77057,3	194764,8	74567,7	181729,3	84654,1	781574,4	181761,5
- Seedlings	lei	57600,0	24320,0	83840,0	16230,4	60544,0	7849,0	250383,4	58228,7
- Fertilizers	lei	68032,0	29696,0	73088,0	47820,8	37888,0	62549,8	319074,6	74203,4
- Pesticides	lei	35059,2	16774,4	37836,8	10516,5	75187,2	9094,4	184468,5	42899,6
- Other materials	lei	8110,1	6266,9	0,0	0,0	8110,1	5161,0	27648,0	6429,8
2. Expenses with mechanical works	lei	197668,9	104309,3	152693,8	127916,7	173793,9	134431,0	890813,8	207166,0

Indicators	U.M	CROP						TOTAL	
		Wheat	Barley	Corn	Sunflower	Soy	Rape	lei/expl.	€/expl.
		ha							
3. Expenditure on irrigation	lei	0,0	0,0	107520,0	0,0	89600,0	0,0	197120,0	45841,9
4. Supply Expenses	lei	8440,1	3852,9	9738,2	3728,4	9086,5	4232,7	39078,7	9088,1
5. Temporary labor expenses	lei	0,0	0,0	0,0	0,0	0,0	0,0	0,0	0,0
6. Insurance	lei	7631,1	3783,0	11071,8	4389,1	9989,4	4540,3	41404,8	9629,0
II. FIXED COSTS	lei	49438,6	25202,5	150911,4	39934,1	102460,9	30140,8	398088,3	92578,7
- Permanent labor expenses	lei	6646,4	3931,0	88873,1	13241,7	45262,5	3698,6	161653,3	37593,8
- Overheads	lei	11446,7	5674,5	16607,7	6583,6	14984,2	6810,5	62107,2	14443,5
- Interest	lei	22387,0	11117,9	32631,1	12943,6	29423,1	13367,0	121869,7	28341,8
- Depreciation	lei	8958,5	4479,1	12799,5	7165,3	12791,1	6264,7	52458,2	12199,6
E. (=) TAXABLE INCOME	lei	100025,2	23800,0	129315,0	51867,6	80146,5	65405,2	450559,5	104781,3
(-) Taxes	lei	0,0	0,0	0,0	0,0	0,0	0,0	0,0	0,0
F. (=) NET REVENUE+SUBSIDIES	lei	168025,2	57800,0	197315,0	106267,6	148146,5	113005,2	790559,5	183851,0
G. RATA TAXABLE INCOME (%)	%	23,2	11,1	20,6	20,7	14,1	25,4	19,2	19,2
H. RATA NET REVENUE+SUBSIDIES	%	38,9	27,0	31,5	42,4	26,1	43,8	33,7	33,7

Source: author's own computation

Appendix 12

Annual Budget for the Holding of 800 Ha, 2013–2017 Lowland Area

INDICATORS	U.M	2012		2013		2014		2015		2016		2017	
		lei/expl.	€/expl.	lei/expl.	€/expl.	lei/expl.	€/expl.	lei/expl.	€/expl.	lei/expl.	€/expl.	lei/expl.	€/expl.
A. PRODUCTICN VALUE	lei	3418400,0	794976,7	3641760,0	846920,9	3894528,0	905704,2	4161984,0	967903,3	4444128,0	1033518,1	4759200,0	1106790,7
A[1]. From the main production	lei	3352000,0	779534,9	3575360,0	831479,1	3828128,0	890262,3	4095584,0	952461,4	4377728,0	1018076,3	4692800,0	1091348,8
B (+) SUBSIDIES	lei	544000,0	126511,6	544000,0	126511,6	544000,0	126511,6	544000,0	126511,6	544000,0	126511,6	544000,0	126511,6
C (=) GROSS PRODUCT	lei	3962400,0	921488,4	4185760,0	973432,6	4438528,0	1032215,8	4705984,0	1094414,9	4988128,0	1160029,8	5303200,0	1233302,3
D (-) TOTAL EXPENSES	lei	2965050,2	689546,5	3113268,8	724016,0	3261531,2	758495,6	3439435,2	799868,7	3617328,0	841239,1	3795216,0	882608,4
D[1]. From the main production	lei	2964983,8	689531,1	3113202,4	724000,6	3261464,8	758480,2	3439368,8	799853,2	3617261,6	841223,6	3795149,6	882592,9
I. VARIABLE COSTS	lei	2461041,2	572335,2	2584093,3	600951,9	2707145,4	629568,7	2854807,8	663908,8	3002470,3	698248,9	3150132,8	732589,0
1. Raw materials and supplies expenditure	lei	991393,6	230556,7	1040963,3	242084,5	1090533,0	253612,3	1150016,6	267445,7	1209500,2	281279,1	1268983,8	295112,5
- Seedlings	lei	312979,2	72785,9	328628,2	76425,2	344277,1	80064,4	363055,9	84431,6	381834,6	88798,7	400613,4	93165,9
- Fertilizers	lei	412347,2	95894,7	432964,6	100689,4	453581,9	105484,2	478322,8	111237,8	503063,6	116991,5	527804,4	122745,2
- Pesticides	lei	230585,6	53624,6	242114,9	56305,8	253644,2	58987,0	267479,3	62204,5	281314,4	65422,0	295149,6	68639,4
- Other materials	lei	35481,6	8251,5	37255,7	8664,1	39029,8	9076,7	41158,7	9571,8	43287,6	10066,9	45416,4	10562,0
2. Expenses with mechanical works	lei	1121382,0	260786,5	1177451,1	273825,8	1233520,2	286865,2	1300803,2	302512,4	1368086,1	318159,6	1435369,0	333806,7

INDICATORS	U.M	2012		2013		2014		2015		2016		2017	
		lei/expl.	€/expl.	lei/expl.	€/expl.	lei/expl.	€/expl.	lei/expl.	€/expl.	lei/expl.	€/expl.	lei/expl.	€/expl.
3. Expenditure on irrigation	le	246400,0	57302,3	258720,0	60167,4	271040,0	63032,6	285824,0	66470,7	300608,0	69908,8	315392,0	73347,0
4. Supply Expenses	le	49569,7	11527,8	52048,2	12104,2	54526,6	12680,6	57500,8	13372,3	60475,0	14064,0	63449,2	14755,6
5. Temporary labor expenses	le	0,0	0,0	0,0	0,0	0,0	0,0	0,0	0,0	0,0	0,0	0,0	0,0
6. Insurance	lei	52295,9	12161,8	54910,7	12769,9	57525,5	13378,0	60663,3	14107,7	63801,0	14837,5	66938,8	15567,2
II. FIXED COSTS	lei	504008,9	117211,4	529175,5	123064,1	554385,8	128926,9	584627,4	135959,9	614857,7	142990,2	645083,2	150019,4
- Permanent labor expenses	lei	206051,4	47918,9	216354,0	50314,9	226656,5	52710,8	239019,6	55586,0	251382,7	58461,1	263745,8	61336,2
- Overheads	lei	78443,9	18242,8	82366,1	19154,9	86288,3	20067,0	90994,9	21161,6	95701,6	22256,2	100408,2	23350,7
- Interest	lei	153913,6	35793,9	161609,3	37583,6	169305,0	39373,3	178539,8	41520,9	187774,6	43668,5	197009,4	45816,1
- Depreciation	lei	65600,0	15255,8	68846,1	16010,7	72136,0	16775,8	76073,0	17691,4	79998,8	18604,4	83919,8	19516,2
E. (=) TAXABLE INCOME	lei	387016,2	90003,8	462157,6	107478,5	566663,2	131782,1	656215,2	152608,2	760466,4	176852,7	897650,4	208755,9
(-) Taxes	lei	0,0	0,0	0,0	0,0	0,0	0,0	0,0	0,0	0,0	0,0	0,0	0,0
F. (=) NET REVENUE + SUBSIDIES	lei	931016,2	216515,4	1006157,6	233990,1	1110663,2	258293,8	1200215,2	279119,8	1304466,4	303364,3	1441650,4	335267,5
G. RATA TAXABLE INCOME	%	13,1	13,1	14,8	14,8	17,4	17,4	19,1	19,1	21,0	21,0	23,7	23,7
H. RATA NET REVENUE + SUBSIDIES	%	31,4	31,4	32,3	32,3	34,1	34,1	34,9	34,9	36,1	36,1	38,0	38,0

Source: author's own computation

Appendix 12a

BUDGET of 800 Ha Farmin Lowland Area
Basic Variant – Year 2012

Indicators	U.M	CROP						TOTAL	
		Wheat	Barley	Corn	Sunflower	Soy	Rape	lei/expl.	€/expl.
		ha							
		160,00	80,00	160,00	128,00	160,00	112,00	800,00	800,00
		20,0%	10,0%	20,0%	16,0%	20,0%	14,0%	100%	100%
A. PRODUCTION VALUE	lei	616320,0	280000,0	976800,0	351360,0	817600,0	376320,0	3418400,0	794976,7
A1. From the main production	lei	608000,0	272000,0	950400,0	345600,0	806400,0	369600,0	3352000,0	779534,9
B (+) SUBSIDIES	lei	108800,0	54400,0	108800,0	87040,0	108800,0	76160,0	544000,0	126511,6
C (=) GROSS PRODUCT	lei	725120,0	334400,0	1085600,0	438400,0	926400,0	452480,0	3962400,0	921488,4
D (-) TOTAL EXPENSES	lei	539976,9	267757,4	802720,8	313173,4	718913,8	322507,8	2965050,2	689546,5
D1. From the main production	lei	539968,6	267749,4	802694,4	313167,7	718902,6	322501,1	2964983,8	689531,1
I. VARIABLE COSTS	lei	478176,8	236253,1	609075,5	263252,4	589460,7	284822,7	2461041,2	572335,2
1. Raw materials and supplies expenditure	lei	211001,6	96321,6	252592,0	93209,6	232451,2	105817,6	991393,6	230556,7
- Seedlings	lei	72000,0	30400,0	104800,0	20288,0	75680,0	9811,2	312979,2	72785,9
- Fertilizers	lei	85040,0	37120,0	100496,0	59776,0	51728,0	78187,2	412347,2	95894,7
- Pesticides	lei	43824,0	20968,0	47296,0	13145,6	93984,0	11368,0	230585,6	53624,6
- Other materials	lei	10137,6	7833,6	0,0	0,0	11059,2	6451,2	35481,6	8251,5
2. Expenses with mechanical works	lei	247086,2	130386,7	195265,1	159895,9	220709,3	168038,8	1121382,0	260786,5
3. Expenditure on irrigation	lei	0,0	0,0	134400,0	0,0	112000,0	0,0	246400,0	57302,3
4. Supply Expenses	lei	10550,1	4816,1	12629,6	4660,5	11622,6	5290,9	49569,7	11527,8

Indicators	U.M	CROP						TOTAL	
		Wheat	Barley	Corn	Sunflower	Soy	Rape	lei/expl.	€/expl.
		ha							
5. Temporary labor expenses	lei	0,0	0,0	0,0	0,0	0,0	0,0	0,0	0,0
6. Insurance	lei	9538,9	4728,8	14188,8	5486,4	12677,7	5675,4	52295,9	12161,8
II. FIXED COSTS	lei	61800,1	31504,3	193645,3	49921,1	129453,1	37685,1	504008,9	117211,4
- Permanent labor expenses	lei	8308,0	4913,8	114553,4	16552,1	57100,9	4623,2	206051,4	47918,9
- Overheads	lei	14308,4	7093,1	21283,2	8229,5	19016,5	8513,1	78443,9	18242,8
- Interest	lei	27983,7	13897,3	41808,6	16179,5	37335,7	16708,8	153913,6	35793,9
- Depreciation	lei	11200,0	5600,0	16000,0	8960,0	16000,0	7840,0	65600,0	15255,8
E. (=) TAXABLE INCOME	lei	68031,4	4250,6	147705,6	32432,3	87497,4	47098,9	387016,2	90003,8
(-) Taxes	lei	0,0	0,0	0,0	0,0	0,0	0,0	0,0	0,0
F. (=) NET REVENUE+SUBSIDIES	lei	176831,4	58650,6	256505,6	119472,3	196297,4	123258,9	931016,2	216515,4
G. RATA TAXABLE INCOME (%)	%	12,6	1,6	18,4	10,4	12,2	14,6	13,1	13,1
H. RATA NET REVENUE+SUBSIDIES (%)	%	32,7	21,9	32,0	38,1	27,3	38,2	31,4	31,4

Source: author's own computation

Appendix 12b

Budget of 800 Ha Holding in Lowland Area
Projected Version–Year 2017

Indicators	U.M	C ROP						TOTAL	
		Wheat	Barley	Corn	Sunflower	Soy	Rape	lei/expl.	€/expl.
		ha							
		160,00	80,00	160,00	128,00	160,00	112,00	800,00	800,00
		20,0%	10,0%	20,0%	16,0%	20,0%	14,0%	100%	100%
A. PRODUCTION VALUE	lei	859520,0	388800,0	1356960,0	489600,0	1140160,0	524160,0	4759200,0	1106790,7
A1. From the main production	lei	851200,0	380800,0	1330560,0	483840,0	1128960,0	517440,0	4692800,0	1091348,8
B (+) SUBSIDIES	lei	108800,0	54400,0	108800,0	87040,0	108800,0	76160,0	544000,0	126511,6
C (=) GROSS PRODUCT	lei	968320,0	443200,0	1465760,0	576640,0	1248960,0	600320,0	5303200,0	1233302,3
D (-) TOTAL EXPENSES	lei	691168,0	342728,0	1027472,0	400857,6	920192,0	412798,4	3795216,0	882608,4
D1. From the main production	lei	691159,7	342720,0	1027445,6	400851,8	920180,8	412791,7	3795149,6	882592,9
I. VARIABLE COSTS	lei	612066,3	302404,0	779616,7	336963,0	754509,8	364573,1	3150132,8	732589,0
1.Raw materials and supplies expenditure	lei	270082,0	123291,6	323317,8	119308,3	297537,5	135446,5	1268983,8	295112,5
- Seedlings	lei	92160,0	38912,0	134144,0	25968,6	96870,4	12558,3	400613,4	93165,9
- Fertilizers	lei	108851,2	47513,6	128634,9	76513,3	66211,8	100079,6	527804,4	122745,2
- Pesticides	lei	56094,7	26839,0	60538,9	16826,4	120299,5	14551,0	295149,6	68639,4
- Other materials	lei	12976,1	10027,0	0,0	0,0	14155,8	8257,5	45416,4	10562,0
2. Expenses with mechanical works	lei	316270,3	166895,0	249939,4	204666,8	282507,9	215089,7	1435369,0	333806,7
3. Expenditure on irrigation	lei	0,0	0,0	172032,0	0,0	143360,0	0,0	315392,0	73347,0
4. Supply Expenses	lei	13504,1	6164,6	16165,9	5965,4	14876,9	6772,3	63449,2	14755,6

Indicators	U.M	C ROP						TOTAL	
		Wheat	Barley	Corn	Sunflower	Soy	Rape	lei/expl.	€/expl.
		ha							
5. Temporary labor expenses	lei	0,0	0,0	0,0	0,0	0,0	0,0	0,0	0,0
6. Insurance	lei	12209,8	6052,8	18161,7	7022,5	16227,4	7264,5	66938,8	15567,2
II. FIXED COSTS	lei	79101,7	40324,0	247855,3	63894,6	165682,2	48225,3	645083,2	150019,4
- Permanent labor expenses	lei	10634,2	6289,7	146628,4	21186,6	73089,1	5917,7	263745,8	61336,2
- Overheads	lei	18314,7	9079,2	27242,5	10533,8	24341,1	10896,8	100408,2	23350,7
- Interest	lei	35819,2	17788,6	53515,1	20709,7	47789,7	21387,2	197009,4	45816,1
- Depreciation	lei	14333,6	7166,5	20469,3	11464,4	20462,3	10023,6	83919,8	19516,2
E. (=) TAXABLE INCOME	lei	160040,3	38080,0	303114,4	82988,2	208779,2	104648,3	897650,4	208755,9
(-) Taxes	lei	0,0	0,0	0,0	0,0	0,0	0,0	0,0	0,0
F. (=) NET REVENUE+SUBSIDIES	lei	268840,3	92480,0	411914,4	170028,2	317579,2	180808,3	1441650,4	335267,5
G. RATA TAXABLE INCOME (%)	%	23,2	11,1	29,5	20,7	22,7	25,4	23,7	23,7
H. RATA NET REVENUE+SUBSIDIES	%	38,9	27,0	40,1	42,4	34,5	43,8	38,0	38,0

Source: author's own computation

www.ingramcontent.com/pod-product-compliance
Ingram Content Group UK Ltd.
Pitfield, Milton Keynes, MK11 3LW, UK
UKHW020118290926
15471UKWH00003B/199

9 783631 658260